Historic Bridges
of Buckinghamshire

Marshall G. Hall

WIND*gather*
PRESS

Windgather Press is an imprint of Oxbow Books

Published in the United Kingdom in 2021 by
OXBOW BOOKS
The Old Music Hall, 106-108 Cowley Road, Oxford, OX4 1JE

and in the United States by
OXBOW BOOKS
1950 Lawrence Road, Havertown, PA 19083

Hardback Edition: ISBN 978-1-91118-892-6
Digital Edition: ISBN 978-1-91118-893-3 (epub)

A CIP record for this book is available from the British Library

Printed in India by Replika Press Pvt. Ltd.

For a complete list of Windgather titles, please contact:

United Kingdom
OXBOW BOOKS
Telephone (01865) 241249
Email: oxbow@oxbowbooks.com
www.oxbowbooks.com

United States of America
OXBOW BOOKS
Telephone (610) 853-9131, Fax (610) 853-9146
Email: queries@casemateacademic.com
www.casemateacademic.com/oxbow

Oxbow Books is part of the Casemate group

Cover images by Marshall Hall
Cover design by Chris Sims

CONTENTS

INTRODUCTION

Historically rivers have been a hub for human settlement and have long been a key part of local livelihoods, history, and culture, as well as playing a role today in providing services and leisure to people who live around them. All four of the earliest human civilisations were formed on great rivers: the Nile, Euphrates, Indus, and the Yellow rivers. Cities have traditionally sprung up at bridgeheads or where a river could be forded at any time of the year. Some examples in the UK are London, Oxford, and Cambridge. The most ancient, vital, and interesting architectural structures linked to the use of these rivers are bridges, and people hold a deep fascination for them. Who isn't captivated by bridges? Who hasn't held their breath as their parents (or they themselves) drove across a particularly high or lengthy span? Bridges have always maintained a lure for people and there is even a specific word for someone who loves bridges; pontist:

A historic bridge enthusiast who enjoys either lobbying for preservation and/or who enjoys visiting and photographing historic bridges.

Perhaps the allure of bridges is their essential nature, their ability to overcome a natural obstacle, their engineering achievement, their pivotal historical significance, and the views afforded by them, but also perhaps because they are a metaphor for so many things in life. Bridges connect two things that nature had no intention of connecting. They represent humans evolving in their relationship with the environment. The metaphorical bridge represents hope, opportunities, potential, and healing, because it is symbolic of an intentional connection only made possible by a mutual understanding between two separate sides, countries, disciplines, or people.

Bridges have long been the subject of superstitions. Because they are a literal passage between two places, perhaps they symbolise the fear of movement from one stage of life to the next. Superstitious rituals are most powerful when there's underlying fear and anxiety involved. In many parts of the world it's considered unlucky to be the first person to cross a new bridge; the belief is that the devil, envious of the human ability to create a structure so complex, will take the soul of the first living creature to make its way across. Workers sometimes leave money in the mortar of a new bridge to protect it and ensure good luck, perhaps a remnant of a rumoured earlier custom involving human sacrifice. Many people lift their feet when driving over a bridge believing that if you don't, you'll never meet your true love or get married. Some drivers even believe that you will die young if you do not observe this driving superstition. Other people touch the roof of their car, or spit, when driving under a bridge while a train or other vehicle is passing overhead, supposedly to make sure it does not collapse, although this seems both ineffective and unhygienic. The reverse of this belief is that you must hold the top of your head while crossing a bridge to keep the bridge from collapsing. While compiling this work I was told that if you say goodbye to a friend on a bridge, you will never see each other again.

Whether allure or superstition, bridges have for centuries been much more than a simple water crossing and, in many cases, became status symbols or political statements. Historically many bridges have been emblazoned with the crests of local families or parishes. Bridges have become the icons of cities and countries. Mention the bridge name and the city springs to mind; the Golden Gate, the Ponte Vechio, the Bridge of Sighs, the Gateshead Millennium Bridge, Tower Bridge, the Forth Bridge, the Henry Hudson Bridge, and many more.

Because of the job they perform, the weight of focused traffic, and where they are located, bridges have always imposed a high financial burden of upkeep responsibility and arguments have flourished throughout history of who was liable for payment. Sometimes the obligation fell to the local parish, occasionally private companies set up tolls to fund maintenance, and sometimes local authorities absorbed the cost. All the bridges in this book, even the very old ones, are replacements for earlier versions.

There are literally thousands of bridges in Buckinghamshire, varying vastly in size, style, and materials. Many are stone, a few are wooden, and there are numerous brick and more modern steel and concrete constructions. The very earliest bridges were simple structures that were built from easily accessible natural resources like wooden logs, stones, and dirt. Consequently, they only had the ability to span very close distances, and their structural integrity was not high. None of these survives in modern Buckinghamshire. Over the millennium the materials and techniques have evolved to produce the wonders of modern bridge engineering we see today, such as the elegant and beautiful Millau Viaduct Bridge over the River Tarn in southern France. The building of such modern bridges was only made possible by the Roman perfection of cement, the cofferdam, and concrete, all of which they brought to England and Buckinghamshire.

We have lost many of our older bridges to the ravages of time and the modern practice of culvertisation and urban development. A few of our older bridges remain though, and their beauty and pivotal role in our history is starting to be recognised.

If looking at a bridge is more exciting than crossing that bridge, then you can be sure that it is a very beautiful bridge!

Mehmet Murat Ildan (born 1965)

This book looks at the wonderful historic bridges that make up the chronology of Buckinghamshire. It is organised by river starting in the north of the county and progressing southwards. A few bridges in the book are identified as feature bridges. These are structures that are of particular historical significance and are architecturally important. The listing for these feature bridges will contain further historical and architectural detail as well as additional photographs.

Of course, there are many bridges in the county that are not included here and if your favourite is missing yet seems to meet the criteria, I apologise. Please let me know and perhaps we can include it in the next edition.

Criteria for Inclusion

- The original structure is more than 100 years old
- The bridge has some historical significance
- It is architecturally and visually interesting
- Its position is within, or forms the border of, Buckinghamshire
- The structure (or remains) is still standing and is publicly accessible
- The bridge spans a river, lake, or stream. No railway or canal bridges are included

About the County of Buckinghamshire

Buckinghamshire is a wildly irregular shaped county formed partly by history and partly by its natural river boundaries. The county's rich natural resources gave rise to many industries in the 19th century. It is considered to be one of the 'Home Counties' and is one of the 'shires' just north of London. The geographic 'traditional' or 'ancient' county takes in the valley of the River Great Ouse in the north, the Vale of Aylesbury and the beautiful Chiltern Hills surrounding Wendover, and the Thames River Valley on its southern border. It also includes the unitary authorities of Milton Keynes, Slough, and parts of Windsor and Maidenhead. The county seat is not the city of Buckingham as one might expect, but Aylesbury.

Historically, Buckinghamshire has been significant with the formation of great estates such as at Waddesdon, Claydon, Stowe, West Wycombe, Hampden House, and many others. Access to London was a factor in the county's development and has continued to be responsible for population growth in the county today. Several of the bridges in this book are associated with these great houses and a number of Civil War battles were waged on Buckinghamshire soil.

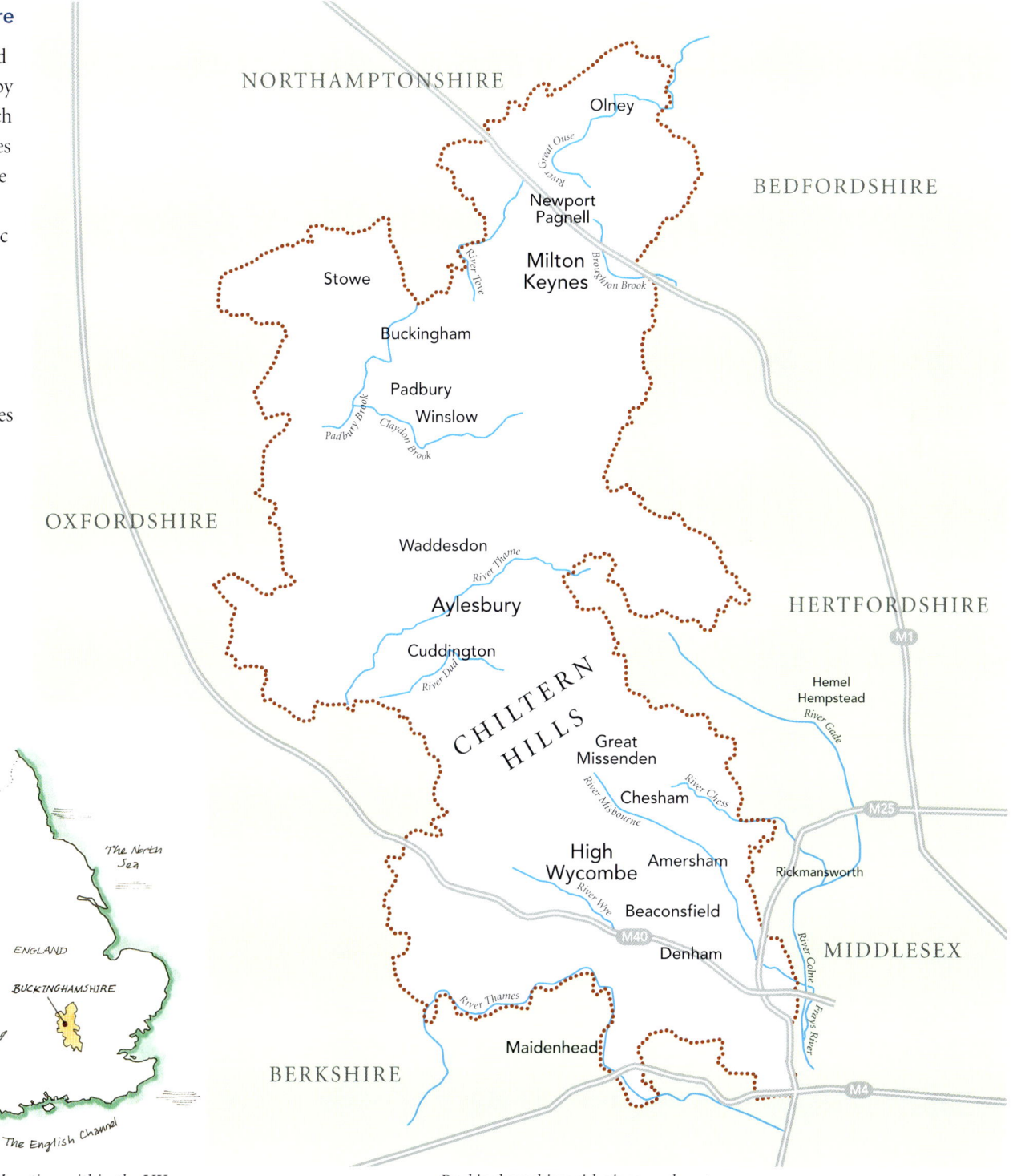

Buckinghamshire location within the UK

Buckinghamshire with rivers and motorways

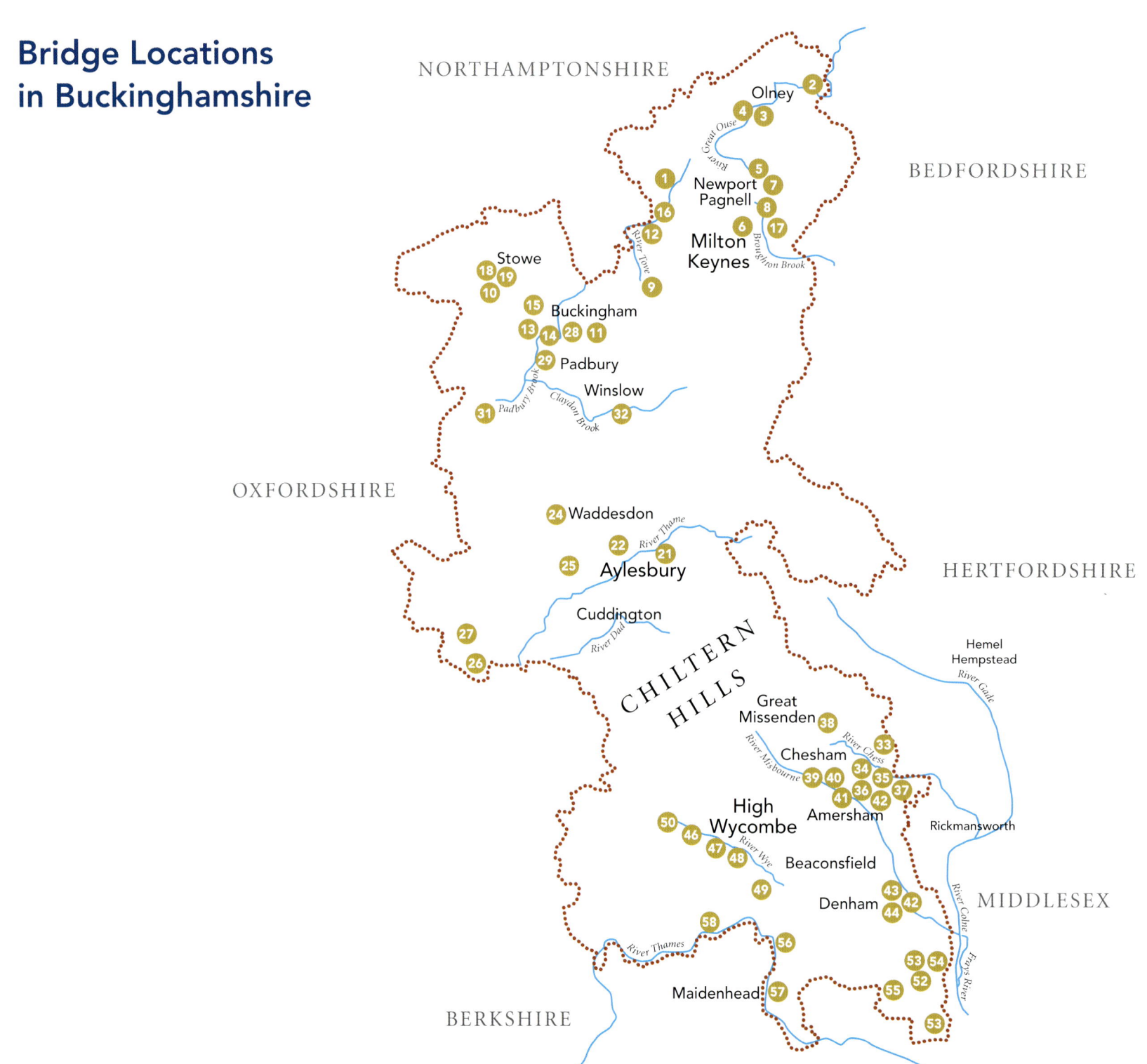

Bridge Locations in Buckinghamshire

NORTHAMPTONSHIRE

BEDFORDSHIRE

OXFORDSHIRE

HERTFORDSHIRE

MIDDLESEX

BERKSHIRE

Olney

Newport Pagnell

Milton Keynes

Stowe

Buckingham

Padbury

Winslow

Waddesdon

Aylesbury

Cuddington

CHILTERN HILLS

Hemel Hempstead

Great Missenden

Chesham

Amersham

High Wycombe

Beaconsfield

Denham

Rickmansworth

Maidenhead

River Great Ouse

River Tove

Broughton Brook

Padbury Brook

Claydon Brook

River Thame

River Dad

River Gade

River Chess

River Misbourne

River Wye

River Thames

River Colne

Frays River

4

1. Castlethorpe Station Road Bridge
2. Turvey Bridge
3. Olney Bridge
4. Goosey Bridge
5. Tyringham Bridge
6. Little Linford Lane Bridge
7. Sherington Road Bridge
8. North Bridge – Newport Pagnell
9. Leckhampstead Bridge
10. Water Stratford Road Bridge
11. Radclive-cum-Chackmore Bridge
12. Old Stratford Bridge
13. London Road Bridge – Buckingham
14. Old Farm Bridge
15. Lord's Bridge
16. Cosgrove Iron Trunk Aqueduct
17. Tickford Iron Bridge
18. Palladian Bridge
19. Oxford Bridge
20. Shell Bridge

21. Holman's Bridge
22. Eythrope Park Bridge
23. Wotton Underwood Bridge
24. Bridgeway Bridge – Cuddington
25. Hartwell House Bridge
26. Thame Bridge
27. Ickford Bridge / Whirlpool Arch Bridge
28. Thornborough Bridge
29. Oxlane Bridge
30. Claydon House Bridge
31. Three Bridges Mill
32. Shipton Brook Bridge
33. Chesham Town Bridge
34. Bois Moor Road Bridge
35. Stoney Lane Bridge
36. Chenies Hill Bridge
37. Chenies Place Woodside
38. Missenden Abbey Bridges
39. Highmore Cottages Bridge
40. Mill Lane Bridge

41. Priory Bridge
42. Village Road Bridge
43. Old Bridge at Denham Place
44. Old Mill Road Bridge
45. West Wycombe Park Bridges
46. Pepperpots Bridge
47. Queen Victoria Bridge
48. Pann Mill Bridge
49. Windsor Hill Bridge
50. Hughenden Manor Bridge
51. Iver Bridge
52. Iver Lane Bridge
53. Thorney Mill Road Bridge
54. Rockingham Road Bridge
55. Repton Bridge
56. Cookham Bridge
57. Maidenhead Bridge
58. Marlow Bridge

Basic Components of a Bridge

There is a glossary at the back of this book which will help you understand the terminology of bridge construction. Like every subject, bridges have their own vocabulary and some of it has been used in the information provided about each bridge.

The components of all bridges, however complex or simple, fall into two large categories: superstructure (those elements above the deck/roadway) and substructure (those below the deck/roadway).

The superstructure is comprised of the slab (carriageway), girders or trusses, etc. The superstructure bears the load passing over the span and transmits the weight caused by the vehicle or pedestrian to the substructure.

The substructure is comprised of foundations, piers, abutments, and wing walls, which are provided to transmit the load of the superstructure to the earth. These are known as bearings.

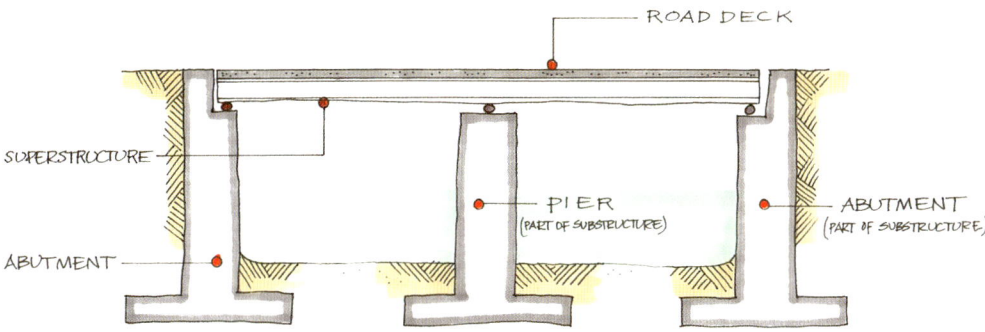

SUBSTRUCTURE AND SUPERSTRUCTURE

The abutments of a bridge alone may not have enough strength to take the load the superstructure places directly on them. This is especially true if the load limit is high or the span is long. To support these loads, bearings such as piers may be constructed to help carry the load from the deck downward and distribute it evenly over the substructure material (earth, concrete, or stone). *(History of Bridges, 2020)*

Rivers in Buckinghamshire

River Chess – Chesham

Colne Brook – Uxbridge Moor

River Colne – London

Frays River – West Drayton

River Gade – Dagnal (no bridges in this book)

River Great Ouse

Jubilee River – Maidenhead (no bridges in this book)

River Lyde – Bledlow (no bridges in this book)

River Misbourne – Great Missenden

River Ouzel / Lovat – Newport Pagnell

River Rau – Quainton Hill (no bridges in this book)

River Thame – Vale of Aylesbury

Padbury Brook – Padbury

River Thames – Henley-on-Thames

River Tove – Cosgrove

River Wraysbury – West Drayton (no bridges in this book)

Hughenden Stream – West Wycombe

River Wye – High Wycombe

Shipton Brook – Winslow

River Dad & Dad Brook – Cuddington

Broughton Brook – Milton Keynes

Claydon Brook – Winslow

Bridge Types Based on Superstructure Construction

There are many ways of classifying bridges into types based on criteria such as materials, function, level crossing, flood level, permanence, etc. However, unless you are an engineer, types of bridges are usually determined by the superstructure. They generally fall into the list below. Some bridges may be combinations of types.

Beam (Girder) Bridge

A beam bridge, also known as a girder bridge, is one of the simplest types of bridge to build. In this form of bridge, the deck/roadway sits on top of horizontal girders, which are supported by abutments at both ends. It is the best design to span short distances that are not more than 76 metres. A beam bridge is usually built with concrete and steel.

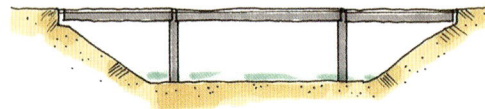

BEAM

Truss Bridge

A truss bridge is like a beam bridge except that the supporting horizontal components are placed to the side of the deck/roadway and are not solid beams. Instead they are made of welded repeating triangular patterns of steel known as trusses, which resist downward load pressure. Given that a triangle cannot be easily distorted by stress, a truss gives a stable form capable of supporting considerable external loads over a large span. Trusses are popular for bridge building because they use a relatively small amount of material for the weight they can support.

The truss bridge was widely used during the Second World War in a modular and transportable format. Sir Donald Coleman Bailey, a British engineer from Yorkshire, designed truss systems for military bridges that could be quickly conveyed and assembled where needed. These became widely known as 'Bailey Bridges'. After the war these became popular as less expensive army surplus alternatives to other bridging options. Buckinghamshire still has several of these bridges.

TRUSS

Arch Bridge

An arch bridge is one of the earliest types of bridges and is by far the most common type of bridge in Buckinghamshire. Arch bridges came into use over 3,000 years ago and remained common until the Industrial Revolution and the invention of advanced materials, enabling architects to create other modern bridge designs. They are very strong structures. The weight does not push straight down on the bridge but instead it's carried outward along the curve of the arch to the supports at both ends. The ends, which are known as abutments, carry the load and keep the arch and the roadway above it in place.

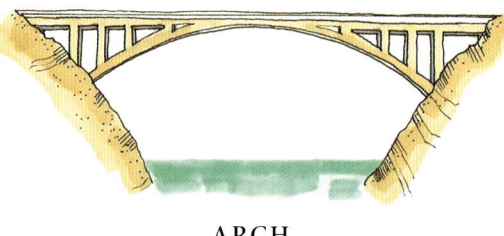

ARCH

Cantilever Bridge

A cantilever bridge uses cantilevers, which are horizontal structural 'arms' projecting out into space supported only on one end. Like a diving board, one arm end is anchored on the shore, the other end of the arm is projected toward the centre. One of these is built on each shore and a section is built at the centre connecting the ends of the two arms. The steel truss cantilever bridge was a major engineering breakthrough when first put into practice by engineer Heinrich Gerber in 1867. Cantilever bridges can span distances of over 460 metres and can be more easily constructed at difficult crossings by virtue of using little or no falsework. The most famous example of a cantilever bridge in the UK is the Forth Bridge outside of Edinburgh.

CANTILEVER

Suspension Bridge

Suspension bridges are longer than other types of bridges and are used to span distances from 610 metres to 2,134 metres. A suspension bridge makes use of huge main supporting cables, which extend from one end of the bridge to the other end. The cables rest on the top of high towers and are securely attached at each end by anchorages. The deck/roadway is 'suspended' in place by the suspension cables that are attached to the large overhead supporting cables. Buckinghamshire has only one suspension bridge crossing the Thames at Marlow.

SUSPENSION

Cable-Stayed Bridge

Cable-stayed bridges are similar to suspension bridges in that they have towers and a deck that is held by cables, but cable-stayed bridges hold the deck in place by connecting it directly to the towers instead of via supporting cables. Cable-stayed bridges are mostly used for medium spans between 152 metres and 853 metres. The cables can also be attached to the towers with different patterns.

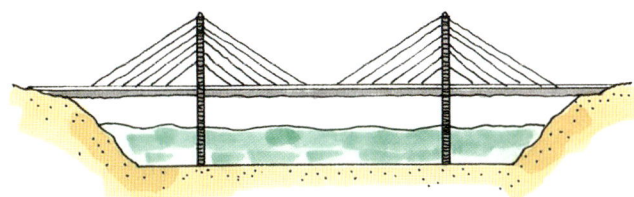

CABLE-STAYED

There are other ways of classifying bridges, but these are the basic types.

In Buckinghamshire we do not have any cantilever or cable-stayed bridges. There are only a couple of truss bridges, but they are both post-Second World War Bailey Bridges and not old enough, so they're not in the book.

Bridge Types Based on Span

Culvert – Less than 6 metres (*Note* may not actually contain a culvert. See glossary*)

Minor – When the bridge span length is between 8 and 30 metres.

Major – A span generally from 30 to 120 metres.

Long span – When the span of a bridge is more than 120 metres. (*History of Bridges, 2020*)

Listed Bridges and Structures

Listing is the term given to the practice of placing buildings, structures, monuments, parklands, gardens, battlefields, and shipwrecks on the list of protected sites. Listing carries with it a specific set of rules safeguarding certain aspects of a site or structure. There are four levels of protection afforded to these structures:

Grade II

Grade II*

Grade I

Monument (*as identified in the 'schedule' of monuments*)

The final decision about listing is taken by the Secretary of State for Digital, Culture, Media, and Sport, but the process, promotion, and enforcement of rules is managed by Historic England. It is an executive body of the British Government.

Historic bridges are commonly split into two categories – Grade I and Grade II listed structures. If a bridge is labelled as a listed structure, you will immediately know that historical value and some form of interest surrounds the structure. There are a number of reasons why a bridge can become listed, but as a general rule, the older a structure is, the more likely it is to be listed. The age and history of the river crossing will play a significant role in whether it is listed, as this can understandably relate to its importance.

Listed bridges will be those that are of special historical, cultural, or architectural interest, or those that are of national importance and have been deemed to be worth protecting. Similarly, every bridge built before 1700, surviving in its original condition, is automatically listed, alongside most bridges built between 1700 and 1840.

What is a Grade I Listed Bridge?

If a bridge is listed as Grade I, this is because the site is of exceptional national, architectural, or historical importance. It is rare to find a Grade I listed bridge compared to a Grade II listed site, simply because of the significance that is placed on such sites. Only 2.5% of all listed bridges in the UK fall into the Grade I category.

What is a Grade II Listed Bridge?

Grade II listed bridges are split into two categories – the majority of which are simply labelled as Grade II, while a small percentage are labelled Grade II*, as they are judged to be of particular national importance or special interest. Just under 6% of all Grade II listed bridges in the UK are listed as Grade II*, with 92% simply labelled as Grade II as they are of special architectural interest. Any bridge or structure erected prior to 1 July 1948 can be listed.

What Does a Listed Bridge Mean?

Listing a bridge is important as it recognises the structure's historical and cultural significance, it also means that in order to make a change to the bridge that could affect its appearance, structure, or historical design, the local authority or bridge owner (if it is privately owned) must apply for listed bridge consent prior to completing any work.

Therefore, it's essential to remember that listed bridges should be celebrated for the rich history and heritage that they represent and that they should be treated with the utmost care and respect, but this doesn't mean that they can't be changed, extended, or repaired.

Scheduled Monuments

A Scheduled Monument is a historic building or site (in this case a bridge) that is included in the Schedule of Monuments kept by the Secretary of State for Digital, Culture, Media, and Sport. The procedure for scheduling is set out in the *Ancient Monuments and Archaeological Areas Act*, 1979.

Applications to schedule or de-schedule a monument are administered by Historic England, who carry out an assessment and make a recommendation to the Secretary of State. The Schedule of Monuments has almost 200,000 entries (2019) and includes sites such as Roman remains, burial mounds, castles, bridges, earthworks, the remains of deserted villages, and industrial sites. Monuments are not graded, but all are, by definition, considered to be of national importance. The protected site of a monument may also include any land adjoining it essential for its support and preservation.

Once a monument is scheduled, any works to it that might affect it, with few exceptions, require scheduled monument consent directly from the Secretary of State (not the local planning authority). Metal detecting on a Scheduled Monument is also illegal without a licence from Historic England.

National Grid Reference Numbers

The National Grid provides a unique reference system that can be applied to all Ordnance Survey (OS) maps of Great Britain, at all scales. Great Britain is covered by grid squares measuring 100 kilometres across and each grid square is identified by two capital letters, i.e. SP, TQ. The two-letter codes can be found printed within the squares in faint blue capitals on Ordnance Survey maps and can also be found in the map key. The first letter, for example 'S', denotes 500 kilometres by 500 kilometres squares and this is subdivided into 25 squares that are 100 kilometres by 100 kilometres within it, making 'ST', 'SU', 'SO' and so on.

There are four main first letters: 'S', 'T', 'N', and 'H' covering Great Britain, plus an 'O' square covering a tiny part of North Yorkshire that is usually below high tide level.

A unique National Grid reference should have this two-letter descriptor followed by the grid reference numbers within that square.

A series of faint blue lines on every map makes up a numbered grid that is used to create the National Grid reference. This is a simple way of finding points and places on a map.

You may find a variety of terms used to describe National Grid references, such as 'OS grid ref', 'grid reference', 'OS map ref', or simply just 'map reference'. While the correct term for these is 'National Grid reference', these terms all mean the same thing, and grid references can be given in a number of different formats.

The numbers going across the map from left to right are called eastings, and go up in value eastwards, and the numbers going up the map from bottom to top are called northings, because they go up in a northward direction.

There are four main types of grid reference:

1. four-figure grid reference, such as '19 45', indicates a 1 kilometre by 1 kilometre square on the map;

2. six-figure grid reference, such as '192 454', indicates a 100 metres by 100 metres square on the map;

3. eight-figure grid reference, such as '1926 4548', indicates a 10 m by 10 m square on the map; and

4. ten-figure grid reference, such as '19267 45487', indicates a 1 metre by 1 metre square on the map.

In practice, it's the six-digit grid reference number that is most commonly used, although the more digits used the more precise the location. GPS devices often specify at least eight-digit grid reference numbers. In this work I have tried to use the ten-digit reference numbers for each bridge wherever they are available.

To use this number with ease, go to **https://gridreferencefinder.com/os.php** and copy & paste the grid reference into their search engine and the location will pop up on the map.

Alphabetical Bridge Listing

A4007 Rockingham Road Bridge – Uxbridge

Bois Moor Road Bridge – Chesham

Bridgeway Bridge – Cuddington

Castlethorpe Station Road Bridge – Castlethorpe

Chenies Hill Bridge – Chenies

Chenies Place Woodside – Chenies

Chesham Town Bridge – Chesham

Claydon House Bridge – Middle Claydon

Cookham Bridge – Cookham and Bourne End

Cosgrove Iron Trunk Aqueduct – Cosgrove

Goosey Bridge – Olney

Eythrope Park Bridge – Stone

Hartwell House Bridge – Aylesbury

Highmore Cottages Bridge – Little Missenden

Holman's Bridge – Aylesbury

Hughenden Manor Bridge – High Wycombe

Ickford Bridge / Whirlpool Arch Bridge – Ickford

Iver Bridge – Iver

Iver Lane Bridge – Iver

Leckhampstead Bridge – Thornborough

Little Linford Lane Bridge – Little Linford

London Road Bridge – Buckingham

Lord's Bridge – Buckingham

Maidenhead Bridge – Maidenhead

Marlow Bridge – Marlow

Mill Lane Bridge – Old Amersham

Missenden Abbey Bridges (2) – Great Missenden

North Bridge – Newport Pagnell

Old Bridge at Denham Place – Denham

Old Farm Bridge – Buckingham

Old Mill Road Bridge – Denham

Old Stratford Bridge – Old Stratford

Olney Bridge – Olney

Oxford Bridge – Stowe

Oxlane Bridge – Padbury

Palladian Bridge – Stowe

Pann Mill Bridge – High Wycombe

Pepperpots Bridge – West Wycombe

Priory Bridge – Old Amersham

Queen Victoria Bridge – High Wycombe

Radclive-cum-Chackmore Bridge – Radclive

Repton Bridge – Stoke Poges

Rockingham Road Bridge – Uxbridge

Shell Bridge – Stowe

Sherington Road Bridge – Sherington

Shipton Brook Bridge – Winslow

Stoney Lane Bridge – Latimer

Thame Bridge – Thame

Three Bridges Mill – Twyford

Tickford Iron Bridge – Newport Pagnell

Thornborough Bridge – Thornborough

Thorney Mill Road Bridge – Thorney

Turvey Bridge – Turvey

Tyringham Bridge – Tyringham

Village Road Bridge – Denham

Water Stratford Road Bridge – Water Stratford

West Wycombe Park Bridges (6) – West Wycombe

Windsor Hill Bridge – Wooburn Green

Wotton Underwood Bridge – Brill

The Bridges

We build too many walls and not enough bridges

Joseph Fort Newton (1880–1950)

North Buckinghamshire Bridges

By far the majority of bridges that meet the criteria for inclusion in this work are in the north of the county. In mid-Buckinghamshire, the Chiltern escarpment runs from the northeast to the southwest of the county. North of this clearly visible geographic feature the land is lower in elevation, gently rolling, and full of sizeable rivers. To the south of the county, below the Chiltern Hills, smaller rivers rise and meander south to the Thames. However, rapid urbanisation in the south of the county has seen the demise of many important bridges.

RIVER TOVE

The River Tove is a tributary of the River Great Ouse. Rising in Northamptonshire about a mile north of Greatworth, it flows for about 15 miles north and east of the town of Towcester (meaning 'camp on the Tove') near Bury Mount, before meeting the Ouse southeast of Cosgrove, north of Milton Keynes. It runs into Buckinghamshire just east of Cosgrove. Its final 5 miles form part of the border between Northamptonshire and Buckinghamshire, running alongside the Grand Union Canal (but the canal is in Northamptonshire) and eventually flowing into the River Great Ouse north of Milton Keynes. There is but one bridge over the Tove in Buckinghamshire.

The three arches viewed from river level of Castlethorpe Station Road Bridge

1 Castlethorpe Station Road Bridge

Bridge Name:	Castlethorpe Station Road Bridge
Location:	South of Castlethorpe on Station Road running between Castlethorpe and Yardley Road north of Milton Keynes.
National Grid Reference:	SP 79185 43960
Crosses:	River Tove
Span:	10.5 metres
Bridge Type:	Three Arch
Materials:	Brick & Stone
Traffic:	Pedestrian & Road
Opened:	Early 19th century
Managed by:	Milton Keynes Council
Historic England Designation:	Unlisted

Information: This little bridge out in the countryside is somewhat overgrown and goes unnoticed as one travels south of Castlethorpe. It is made of local stone and brick forming three arches on two central piers that have newer red brick cutwaters on the north side. The parapets are of the same stone with stone coping. The roadway is approximately 7 metres

Arch detail showing the north abutment

Castlethorpe Station Road Bridge

wide without accommodation for pedestrians. A long causeway crossing a floodplain extends ¼ mile further south with several flood arches. This bridge forms the border between Buckinghamshire and Northamptonshire.

Brick cutwaters on the west side of the Castlethorpe bridge

The Castlethorpe countryside

RIVER GREAT OUSE

There are three rivers in England called the 'Ouse'. This one is situated in the Home Counties, but there are also ones in Yorkshire and Sussex. The Great Ouse rises in the hills between Banbury in Oxfordshire and Brackley in Northamptonshire and then, unusually, it turns northeast to Stony Stratford in Buckinghamshire and is joined by the smaller rivers Tove and Ouzel. It separates Old Stratford and Stony Stratford and runs out of the county at Turvey. Historically, the river has been a major navigation route – one that was recognised by Daniel Defoe in his *A Tour Through the Whole Island of Great Britain*. The River Great Ouse is 150 miles long making it the major navigable river in East Anglia. It is the fourth longest river in the United Kingdom and flows through five counties: Northamptonshire, Buckinghamshire, Bedfordshire, Cambridgeshire, and Norfolk.

② Turvey Bridge – FEATURE BRIDGE

Bridge Name:	**Turvey Bridge**
Location:	**Turvey**
National Grid Reference:	**SP 93785 52420**
Crosses:	**River Great Ouse**
Span:	**209 metres with causeway and abutments**
Bridge Type:	**Multiple Arch, widened in the 20th century**
Materials:	**Stone**
Traffic:	**Pedestrian & Road**
Opened:	**1138**
Managed by:	**Transport for Buckinghamshire jointly with Transport for Bedfordshire**
Historic England Designation:	**Grade I Listed and Scheduled Monument**

Information: Turvey Bridge, spanning the River Great Ouse, is the distinctive entrance to the village of Turvey from the west, and forms the boundary between Bedfordshire and Buckinghamshire. It was designated as a Scheduled Monument in 1930, a nationally important historic building and given protection against unauthorised change.

Turvey Bridge

Like many bridges the approach from the west spans a great flood plain and a causeway of 209 metres long and 10 metres wide gently raises the traveller towards the river crossing. The River Great Ouse splits somewhere north of the village and creates two arms necessitating two bridges. The first, westernmost, bridge is the smaller of the two but is significant in its own right. The substructure consists of four arches, the two central ones of light golden coloured dressed stone spanning 11.5 metres. There is a central supporting pier with a rounded cutwater. The parapets of both bridges are of smaller local stone with standing stone caps.

The easternmost bridge that enters directly into the village is of different, more ancient construction. An argument could be made that this bridge sits mostly within Bedfordshire. It is made of the smaller local stone and is formed of seven arches spanning the east channel of the river (nearest the village) with a raised causeway connecting both bridges. Six of the arches on the east bridge have pointed cutwaters on the upstream (south) side of the bridge that are carried up through the parapets of the superstructure creating passing places for pedestrians. The northern side of this bridge is very overgrown.

The earliest record of the bridge is in 1136/38 when the Drayton Charters records that Sampson Fortis granted '2 acres of meadow next Turveie Bridge' to Harrold Priory. Another reference is in 1272 when justices travelling from Northampton to Bedford stopped at the bridge and heard four cases.

By 1402 there are records of a 'hermitage and chapel' built on the bridge, although it had disappeared by the time of the Chantry survey of 1546. It was probable that income from alms funded repair work.

In the 16th century, money was left in the wills of local noblemen, including John, first Lord Mordaunt, towards the repair of the bridge. His will of 1560 records:

> *I will and bequeath to the inhabitants of the Town of Turvey for and towards the new repairing of Turvey bridge, as much as is within the County of Bedford, £40, and for the repairing and amending of the long bridge (the causeway) within the County of Buckingham, to the reparations and keeping whereof the Towns of Hardemeade, Astwood, Lavendon, Newton Blosmafeld and Brafeelde are contributors, £26.12s.3d.*

(Goodland, 2019)

The medieval bridge had two components: a foot causeway and the river bridge. The foot causeway on the west side of the river is shown in maps from 1783; according to John Higgins of Turvey Abbey it had 22 planked and two stone arches and was just over 2 metres wide. There was a raised stone footpath on the north side of the road on the Turvey side of the bridge.

A survey in around 1630 described the bridge as having 'four high arches' over the east channel of the river, and two over the (then) much smaller channel to the west. At this time, the width of the bridge was about 4.25 metres.

The extensive stone parapets that line the causeway

Detail of the stone arches of the western bridge

Turvey Bridge in 1930 before widening (© The Turvey website)

The Bedfordshire county line marker

Bridge parapet with old graffiti

The western bridge just west of the main Turvey Bridge

The Bridge over the Ouse as seen from the Mill 1840
(Watercolour by John Higgins © J. Longuet Higgins)

The bridge in 1839
(Watercolour by John Higgins, © J. Longuet Higgins)

Turvey Bridge with two figures, one fishing
(Watercolour by John Higgins, 1835, © J. Longuet Higgins)

Turvey Bridge Man in Punt 1800–1840
(Watercolour by John Higgins, © J. Longuet Higgins)

Turvey Bridge and the Three Fishes Public House 1820 (The Turvey Website, 2016)

Turvey Bridge carriageway

The bridge looking eastward into Turvey

It was widened in 1920 but by 1930 road traffic had again increased considerably, and the *Bedfordshire Times and Independent* of 13 June reported 'For some years past, however, it has proved quite inadequate to meet modern demands, for there are places in its length where two big lorries are unable to pass one another. The bridge's ancient stones are in many places rudely scraped and scored as a result of unfortunate encounters with motor vehicles.' *(Goodland, 2019)* Again in 1932 it was widened to accommodate the increase in traffic. Reinforced concrete and 25 cm of limestone were added.

3 Olney Bridge

Bridge Name:	Olney Bridge
Location:	A509 just coming into Olney from the south.
National Grid Reference:	SP 88815 50885
Crosses:	River Great Ouse
Span:	25 metres
Bridge Type:	Multiple Arch
Materials:	Dressed Stone
Traffic:	Pedestrian & Road
Opened:	1334 / 1619 / 1832
Managed by:	Milton Keynes Council
Historic England Designation:	Grade II Listed

Information: Olney town is entered from the south via the appropriately named Bridge Street, encompassing several bridges and a causeway collectively called Olney Bridge. The River Great Ouse had originally been crossed at a ford at this point until the reign of Queen Anne, when, in around 1703, a bridge of 'wearisome but necessary length' was built across the whole valley, thus making communication and trade with the south possible throughout the year, even when the river was in flood. The bridge was much dilapidated and rebuilt in 1832. At its northern end this bridge joined the more ancient one said to have been built in 1619, and itself the successor of a bridge which was out of repair in 1334. Near the five old arches an iron bridge (photo) was built in 1894. *(Olney & District Historical Society, 2019)*

Today's bridge was constructed in 1832 and consists of three large arches with a span of 10 metres each, and two smaller arches with a span of around 7 metres each. All are made of a light-coloured dressed sandstone. There are cutwaters on both

The famous Olney Church steeple

19

sides of the bridge that are carried up through the parapets and form very small passing places for pedestrians. The parapets above the stringer course have been assembled separately of smaller stone and the inside of them has been lined with odd modern concrete cladding, presumably for protection. The parapets have prefabricated concrete coping. The roadway is approximately 10.5 metres wide with pedestrian walkways on both sides.

There are three dry flood arches to the south of the main bridge for allowing flood waters to clear the flood plain.

The bridge is probably most well-known for the Battle of Olney Bridge, a skirmish on the bridge in the First English Civil War just outside Olney on 4 November 1643, in which Royalist forces attacked the Parliamentarian forces holding Olney Bridge. Parliament held Newport Pagnell, and Olney was one

The two main arches of Olney Bridge

Re-enactment of the Battle of Olney Bridge during the 1931 Olney Pageant (© Buckinghamshire Archives, 2020)

The three flood arches to the south of the main bridge

Olney Bridge as it looked in 1990. The five arches can be clearly seen in this photo but extensive overgrowth obscures all but the central two today (© Buckinghamshire Archives, 2020)

The ornamental supports for the parapets

The concrete interior facing of the parapets

Historical etching of Olney Bridge

Olney Bridge looking east with the church steeple

of its outposts. Prince Rupert held Northampton for the King and marched on Olney intending to continue to Newport Pagnell. Prince Rupert and his troops took the Olney forces by surprise and the Parliamentarians retreated to the bridge where they made a defiant stand. The Royalists could have won decisively, had it not been for a rumour that Cromwell's reinforcements were seen coming from Newport Pagnell. The Parliamentarian forces held the bridge, and the remaining Royalists retreated.

The Battle of Olney Bridge reputedly left 40 dead with many wounded. Several military artefacts believed to be associated with this skirmish have been found near the bridge. The finds include a few musket balls, and a Civil War sword retrieved by workmen excavating the

Battle of Olney Bridge memorial

foundations of the new iron railed bridge in the 19th century. A row of graves was unearthed in excavations at Emberton Park; these have been tentatively interpreted as casualties of the Olney Bridge skirmish.

The bridge for which the battle is named remains to this day. A memorial to the dead can be found on the site. *(British History Online – Olney, 2001)*

Olney Bridge west side in the 1960s
(© Buckinghamshire Archives, 2020)

Detail of the Olney Bridge voussoirs and keystones

The Iron Bridge at Olney built in 1894 (since replaced)
(© Buckinghamshire Archives, 2020)

4 Goosey Bridge

Bridge Name:	**Goosey Bridge**
Location:	**Emberton Country Park, south side of Olney.**
National Grid Reference:	**SP 88305 50960**
Crosses:	**River Great Ouse**
Span:	**3 metres**
Bridge Type:	**Single Arch**
Materials:	**Stone**
Traffic:	**Pedestrians**
Opened:	**1796**
Managed by:	**Milton Keynes Council**
Historic England Designation:	**Unlisted**

Goosey Bridge upturned capping stones

Goosey Bridge

22

Information: Goosey Bridge, which dates from 1796, links Goosey Island and, indirectly, Little Goosey Island, to the north bank of the River Great Ouse. The name of the islands and bridge are thought to be derivations of the name of the river. All lie within Emberton Country Park just south of Olney and are managed by Milton Keynes Council.

The carriageway at Goosey Bridge

This small but elegant bridge is constructed of local stone without parapets. A small causeway on each end forms the abutments. The deck was originally earth but has now been reinforced with modern concrete, and each side of the carriageway has been coped with upturned stone of the same kind. The arch is constructed of red brick.

Oliver Ratcliff gives the following legend concerning this pond in his *Olney Bucks* Almanack (1907).

> At the north end of the town there is a pond known as the Whirly Pit. This was supposed to be bottomless and to be fed by some mysterious spring. It is a curious fact that it never shows any signs of becoming dry. It contains numbers of carp, and it is very probable it was the fishpond to the old castle or monastery that is supposed to have stood in the vicinity. A remarkable story was current which connects the Whirly Pit with Sway Gog, a meadow some distance away in the direction of Weston. One night the Devil was supposed to have approached Olney by the Warrington Road, in his chariot drawn by four headless horses. The coachmen were also without heads, while to complete the weird details, the night was dark and the hour that of midnight.

> On nearing the town, the coachmen drove straight into the Whirly Pit and continued the journey underground by means of a passage extending as far as Goosey Bridge. Here they emerged into the open with such violence that the meadow was seriously disturbed. And even now, if any person stands astride on this meadow, it is said to sway, as if shuddering at the recollection of that fearful night. (Ratcliff, 1907)

5 Tyringham Bridge – FEATURE BRIDGE

Bridge Name:	Tyringham Bridge
Location:	Tyringham just off the B526 north of Newport Pagnell.
National Grid Reference:	SP 85798 46526
Crosses:	River Great Ouse
Span:	21 metres
Bridge Type:	Single Arch
Materials:	Stone
Traffic:	Pedestrian & Road
Opened:	1793
Managed by:	Milton Keynes Council
Historic England Designation:	Scheduled Monument and Grade I listed

Information: Located on the Tyringham Estate, this bridge has been the subject of at least one historical painting. Tyringham Manor and Estate was held by the Tyringham family throughout the Middle Ages and beyond.

Tyringham Bridge – the western side

The estate has early 20th-century formal gardens situated in a late 18th-century landscape park. At its most extensive the park covered about 100 hectares. The present grounds around the house occupy 12 hectares.

The main approach to the Tyringham Estate is from the south, off the B526, via Filgrave Lane, which leads under Sir John Soane's austere stone gateway of 1794, which is Grade I listed. This stone structure acts as an entrance screen with an archway and flanking lodges. The gateway is set back 30 metres off the road. The lane runs north from the gateway, straight for 100 metres, to cross this elegant single span stone bridge of 1793. It is Grade I listed, and a Scheduled Monument consisting of a single, elegant arch over the River Great Ouse, also designed by Sir John Soane. Typical of Soane is the simple incised decoration between arch and parapet.

The span is covered by a segmental arch of ashlar stone. At road level there is a panelled segmental parapet with coping that has evidence of repair. Steeply rising abutments at each end of the arch are slightly projecting with semi-circular headed niches and plain approach walls with similar parapets either side of the river.

Tyringham Bridge deck/roadway looking south towards the entrance arch

![Tyringham Bridge in the landscape]

Tyringham Bridge in the landscape

Architect – Tyringham Bridge

Sir John Soane was notable for his original, highly personal interpretations of the Neoclassical style. He is considered one of the most inventive architects of his time.

Sir John Soane
(1753–1837)

In 1772 Soane attended the Royal Academy of Arts and was granted a travelling scholarship by King George III. His 'Grand Tour' was to Italy in 1778, returning to England in 1780. During the next few years, he erected many country houses, the designs for which he published in a volume in 1788. As a country house architect, Soane had modest success until he was appointed architect to the Bank of England in 1788. Various government appointments followed, and in 1806 he became Professor of Architecture at the Royal Academy. He was knighted in 1831.

Soane's style is characterised by a tendency to reduce classical elements of design to their structural essentials, the substitution of linear for modelled ornamentation, frequent use of shallow domes and top lighting, and ingenious handling of interior space.

The list of his works is extensive. Some of the finest are his rebuilding of the Bank of England (1788–1833 – later rebuilt) and Tyringham Bridge.

However, he may be best known today for the museum he created in his home. For the benefit of his pupils and other students, Soane began to form collections of antiquities, books, and works of art. Over the years, especially towards the end of his life, he expended large sums of money. In 1824, he purchased the celebrated alabaster sarcophagus brought from Egypt by Belzoni. He acquired Hogarth's two series of pictures, *The Rake's Progress* in 1802, and *The Election* (from Garrick's collection) in 1823, as well as Reynolds' *Snake in the Grass* and a number of good works by the leading painters and sculptors of the day. These, together with many casts and models of the remains of antiquity, gems, rare books, and illuminated manuscripts, and the whole of his own architectural designs, he arranged in his house in 15 Lincoln's Inn Fields, which he transformed into a museum, employing many ingenious devices for economising space.

The museum is today an eclectic collection of many art forms stuffed into a London townhouse. Well worth a visit. **https://www.soane.org**

Image public domain

Tyringham Bridge

Pen and ink and watercolour on paper by John Piper, The Bridge, Tyringham, from the Recording Britain Collection (Buckinghamshire), 1940. Signed and dated (Piper, 1940)

The east wing wall showing evidence of replaced stone

Tyringham Bridge in winter

The approach from the north

Bridge Name:	Little Linford Lane Bridge
Location:	West side of Newport Pagnell. Follow Wolverton Road over the M1 towards Little Linford.
National Grid Reference:	SP 85215 43670
Crosses:	River Great Ouse
Span:	31 metres
Bridge Type:	Arch
Materials:	Brick
Traffic:	Pedestrian & Road
Opened:	1811
Managed by:	Milton Keynes Council
Historic England Designation:	Unlisted

Information: A narrow humpback bridge over the River Great Ouse just yards from the noise of the M1 services at Newport Pagnell. This delightful red brick bridge sits in a rural setting on the road to Little Linford and has a span of over 30 metres covered by two arches made from four layers of the same red brick. The deck/roadway is only 4 metres wide and only one vehicle may cross at a time. There is a small cutwater on the east side and there is evidence that the bridge has had to be repaired frequently over the decades.

Little Linford Lane Bridge looking westward

The carriageway and parapets of Little Linford Lane Bridge

The west side of Little Linford Lane Bridge

Bridge Name:	Sherington Road Bridge
Location:	Sherington Road just south of Sherington hamlet and northeast of Newport Pagnell
National Grid Reference:	SP 88415 45360
Crosses:	River Great Ouse
Span:	60 metres
Bridge Type:	Multiple Arch, widened in the 20th century
Materials:	Stone
Traffic:	Pedestrian & Road
Opened:	1700s (1818 current bridge)
Managed by:	Milton Keynes Council
Historic England Designation:	Unlisted

The substructure of Sherington Bridge showing the 1970s widening with concrete 'wings'

Information: Sherington Road Bridge sits to the north side of the hamlet of Sherington on the road towards Newport Pagnell. The first reference to the bridge was in documents dated from the 13th century but an exact build date is not known. The original bridge had a stone rubble substructure and a wooden superstructure. The upkeep of Sherington Bridge was a continuous problem and in 1815 the two trusts responsible for its maintenance brought a bill of indictment for it to be taken over by the County at the Aylesbury Assizes. The appeal to have the bridge rebuilt at public expense was successful and the new bridge opened in 1818. It has been repaired several times since then, the most recent repair was a widening in the early 1970s with the addition of concrete extension wings visible in the photos.

As this was one of the major crossing points across the River Great Ouse, the passage across the bridge was subject to a toll between 1785–1830. In fact, there were toll gates all around the village between 1753 and 1878. A toll house was erected on the site where Bridge House now stands. In 1830 the toll gate was moved to just north of the waypoint on the Bedford Road.

Sherington Bridge in 1920
(© Buckinghamshire Archives, 2020)

Sherington Bridge from across the reed beds

The rounded cutwaters

Detail of the stone arches with keystone

Sherington Bridge as seen from the river

8 North Bridge – Newport Pagnell

Bridge Name:	North Bridge
Other Names:	South Bridge and Middle Bridge
Location:	B526 Road, High Street, heading north out of Newport Pagnell.
National Grid Reference:	SP 87805 44250
Crosses:	River Great Ouse
Span:	60 metres
Bridge Type:	Multiple Arch
Materials:	Brick and Stone
Traffic:	Pedestrian & Road
Opened:	1809
Managed by:	Milton Keynes Council
Historic England Designation:	Unlisted

Information: Newport Pagnell's North Bridge is actually a series of bridges of differing dates, construction techniques, and materials connected by a raised causeway to traverse a flood plain. Cattle graze this plain most months of the year. Officially locals may call the two bridges the North Bridge and the South Bridge but collectively they are known as North Bridge because they are on the north side of town, and all are connected by a common deck/roadway with a visually pleasing patchwork of brick and stone parapets.

Timber bridges of various forms have been recorded on this site as early as 1311 and their upkeep was a constant drain on resources of the area. In 1809 an Act of Parliament was obtained for their rebuilding, and the present stone bridge over the Ouse replaced the earlier structures. The current bridge

Remains of the 1380 bridge in Ousebank Gardens

(southernmost bridge) crossing the River Great Ouse is a single clear span arch of 15 metres constructed of yellow ashlar stone. The abutments are of the same stone. There are remains of the much earlier original 1380 bridge in the adjoining Ousebank Gardens. In 2009 the bridge celebrated its bicentennial.

The toll house as it looks today

There is an original toll house at the north end of this bridge (now a private residence) with the date of 1809 in the entrance door lintel. The front bay window of the toll house was used as a lookout so that no potential toll collection was missed. Collecting tolls on the bridge was eventually stopped, so the toll house was no longer needed. Instead of demolishing it, it was sold to a Miss Beaty on condition it was only to

Date above the toll house doorway

be used as a private dwelling and not a business. The next owner, William Bateman Bull, was allowed to make an opening in the North Bridge parapet wall near the house to make a road into the field beside it. This can still be seen.

Moving northward, after approximately 100 metres of causeway, another later brick bridge (North Bridge – sometimes also called Middle Bridge and Lathbury Bridge) continues the original bridge northward and there is evidence of several additions and repairs of differing brick colours and styles.

The North Bridge – Newport Pagnell (Wombwell, 1861) (© Public Domain, held by Newport Pagnel Historical Society)

This bridge of three brick arches sits on stone cutwaters.

Further northward are several flood arches that are also constructed of brick and have cutwaters indicating the occasional need to move vast amounts of overflow river water.

North Bridge has been the subject of at least one historic oil painting by

Edwin Wombwell in 1861. Edwin Edward Wombwell was born in 1802 at Stoke Newington, Middlesex and by the 1850s was residing unmarried at Newport Pagnell when he was described as a painter and restorer. He only ever took board and lodgings and in 1861 lived at Wolverton at the Plough Inn, then a widower and artist. Wombwell returned to Newport Pagnell by the 1880s, then calling himself a landscape artist and known locally as 'George Wombwell'.

The northernmost bridge in the 1950s. This is now primarily used as flood relief

The Middle Bridge – Newport Pagnell

Detail of the arch – North Bridge – Newport Pagnell (this is actually the southernmost bridge of three that are collectively known as 'North Bridge')

View of Middle Bridge and South Bridge across the pasture fields 1989 (© Buckinghamshire Archives, 2020)

A patchwork of construction materials and methods make up the three bridges, as shown in their parapets

North Bridge – view along the capping stones towards the toll house

North Bridge – east side

North Bridge – looking across the bridge towards Newport Pagnell (© Buckinghamshire Council Archive)

North Bridge from Ousebank Gardens 1950 (© Buckinghamshire Council Archive)

North Bridge and the toll house 1990 (© Buckinghamshire Council Archive)

9 Leckhampstead Bridge

Bridge Name:	Leckhampstead Bridge
Location:	Thornborough – about 3 miles northeast of Buckingham. Follow the A422 towards Milton Keynes for a couple of miles north to second road on right. Turn right and follow towards Thornborough, past Leckhampstead Wharf House to bridge.
National Grid Reference:	SP 73841 35493
Crosses:	River Great Ouse
Span:	12 metres
Bridge Type:	Four Arch
Materials:	Brick
Traffic:	Pedestrian & Road
Opened:	18th century (Reconstructed in the 1950s)
Managed by:	Transport for Buckinghamshire
Historic England Designation:	Grade II Listed (at risk)

Information: Although the name would place this bridge in the village of Leckhampstead, it is actually closer to Thornborough. This is an unassuming low-profile, 18th-century bridge of older rubble fill cased in English bond brick. There are four arches, the three 2.5 metre central arches carrying the primary river flow and an outer arch on the north side acting as a flood arch. Unusually the four arches are pointed rather than circular and sit on three piers with concrete foundations.

Leckhampstead Bridge

There is a very low parapet, the result of a modern widening effort with concrete coping and later wood posts and railings added. Photographs from the 1950s show different wooden railings without the concrete coping, indicating that there may never have been higher parapets. The carriageway can carry only one vehicle at a time at only just over 4 metres. The west face has two large brick cutwaters. The southern two arches have, for uncertain reasons, off-set 'stepped' brickwork on both sides. However, the evidence of two styles of brickwork and, in the central arch, an upper arch reinforcing a lower arch is an indication that perhaps these sections had been rebuilt. On the east side of the bridge there are stepped corbels below the parapet.

Detail of the 'arch over an arch' reconstruction

Leckhampstead Bridge

Detail of the stepped brickwork on the southern arches

The carriageway

Detail of the unusual pointed arches

33

Bridge Name:	Water Stratford Road Bridge
Other Names:	Bear Bridge
Location:	Water Stratford, west side of Buckingham, south of the village.
National Grid Reference:	SP 65235 34090
Crosses:	River Great Ouse
Span:	31 metres
Bridge Type:	Single Arch
Materials:	Stone
Traffic:	Pedestrian & Road
Opened:	1881
Managed by:	Transport for Buckinghamshire
Historic England Designation:	Unlisted

Information: Water Stratford is a historic village west of Buckingham with Roman roads and is entered from the south via Water Stratford Road. This ancient and picturesque little humpback bridge is about ¼ mile from the village and constructed of local stone. There are short parapets of the same stone with a narrow pedestrian walkway. The abutments and wing walls form a short causeway on both ends of the bridge and are more modern than the bridge itself. There is a stone flood arch about 10 metres to the north of the bridge. The bridge is surrounded by open countryside and pasture. There are modern pipe handrails on both sides of the bridge, which are distinctly out of character for the structure.

Water Stratford Road Bridge southeast side

Water Stratford Road Bridge

The stone flood arch

The low humpback parapets and the narrow walkway

Bridge Name: Radclive-cum-Chackmore Bridge

Location: Just over 1 mile west of Buckingham off the A421 Tingewick Road. Turn north on Radclive Road and follow until it crosses the bridge.

National Grid Reference: SP 67803 33900

Crosses: River Great Ouse

Span: 11.5 metres

Bridge Type: Four Arch

Materials: Brick

Traffic: Pedestrian & Road

Opened: Early to mid-18th century AD 1700–1799

Managed by: Transport for Buckinghamshire

Historic England Designation: Grade II Listed

Radclive-cum-Chackmore Bridge

Information: The villages of Radclive and Chackmore are located just west of Buckingham off the A421. Radclive, historically Radcliffe, or Ratliff, is said to derive its name from the colour of the soil in the area, and an abrupt outcropping near the course of the Ouse.

This fine red brick bridge has four identical three-layer brick voussoir arches, each with a span of 3 metres. These sit on three brick piers with sharp pointed cutwaters. The piers sit on more recent concrete reinforcement at the level of the riverbed. The brickwork continues upwards past the string course to form parapets with stone coping. The bridge has the reinforcing tie rods and retaining plates typical of the era. The deck/roadway is a narrow 4.5 metres allowing only single lane traffic. The bridge sits in a small, picturesque valley leading into the village. There are two slightly smaller flood arches to the south of the main bridge, constructed of the same materials and methods. A field stone causeway wall leads up to the bridge.

Radclive-cum-Chackmore Bridge, view across the valley

Detail of the Radclive-cum-Chackmore flood arches

The Radclive-cum-Chackmore flood arches

Radclive-cum-Chackmore Bridge in 1970 (© Transport for Buckinghamshire)

Radclive-cum-Chackmore Bridge carriageway

Radclive-cum-Chackmore Bridge coping stones

Bridge Name:	Old Stratford Bridge
Other Names:	Stony Stratford Bridge
Location:	Old Stratford, northwest side of Milton Keynes, Watling Street.
National Grid Reference:	SP 78115 40990
Crosses:	River Great Ouse
Span:	31 metres
Bridge Type:	Three Arch
Materials:	Stone
Traffic:	Pedestrian & Road
Opened:	25 July 1835
Managed by:	Milton Keynes Council
Historic England Designation:	Grade II Listed

Information: Old Stratford Bridge is a fine stone bridge over the River Great Ouse, which for several miles forms the boundary between Northamptonshire and Buckinghamshire. It is also known as Stony Stratford Bridge because it connects the towns of Old Stratford and Stony Stratford. 'Stratford' is the Saxon word for 'street over the ford'. This bridge at Old Stratford forms part of the ancient route of Watling Street, which travels all the way to Canterbury.

The original river crossing from which both Old Stratford and Stony Stratford take their names can be identified immediately upstream from the current bridge. The river was made shallower at this point by dropping stones into it, which reinforced the riverbed. This could partly be the reason why it was called 'Stony' Stratford. The ford must have been replaced by a wooden bridge at some point since a wooden bridge was mentioned in 13th-century documents. Tolls were apparently taken at the bridge in the Middle Ages. *(MK Heritage, n.d.)*

In the early 17th century, the bridge crossed the Ouse itself by a single span, flanked on the Buckinghamshire bank by a causeway consisting of three groups of three arches, apparently built to carry water off the meadowland alongside the river. The bridge is said to have been partly destroyed in the Civil War and then to have become dilapidated by the early 19th century. Accidents were common.

On Sunday evening, between nine and ten o'clock, as one Baldwin, a shoemaker, of Stony Stratford, was crossing the bridge to Old Stratford, owing to the darkness of the night he slipped off the causeway into the water, and was unfortunately drowned.

(The Northampton Mercury, 1800)

On 8 November 1823, the year of the great flood, a stagecoach overturned, and the passengers were only just saved. The old bridge finally collapsed in 1833 when a team of horse-drawn wagons were transporting to Coventry pieces of heavy machinery for making plate glass. The bridge at that time was made of timber, and as the last wagon went over the bridge, the timber weakened, and the wagon fell through into the river River Great Ouse. It took 40 harnessed horses to drag it up the steep hill to Old Stratford.

Eventually, in 1834, Parliament passed an act for a new bridge. The level of both the road and the bridge were raised to help prevent flooding. *(MK Heritage, n.d.)*

The repair and maintenance of the bridge had, since the early 16th century, been the responsibility of a local charity, but this was evidently insufficient. In 1834 the two counties that shared responsibility for the bridge provided for the cost of building the new bridge, and its ongoing maintenance was to be divided between Northamptonshire and Buckinghamshire. To facilitate this, tolls were to be collected for 21 years and the charity that had previously maintained the old bridge was to be discharged from this responsibility.

Old Stratford Bridge and the River Great Ouse

That your memorialists submit that such Toll House when inhabited affords a protection to the Old Stratford Bridge and the walls of the causeways thereof and: be the means of preventing injury being done to the same.

That your memorialists beg to represent that there is frequent traffic over the said Bridge by night as well as by day and that the distance from the end of the Town of Stony Stratford to the commencement of Old Stratford is about a quarter of a mile which if the said Toll House be pulled down will be very unsafe for parties travelling along the same especially females.

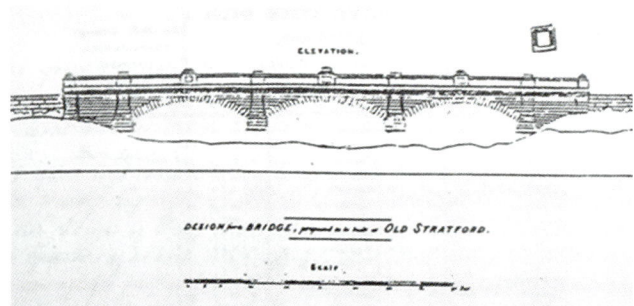
Early architectural drawing of the current bridge at Old Stratford

The original 1857 petition against removing the toll house from the bridge

The extensive causeway wall running south

A very early drawing of the stone bridge at Old Stratford (now replaced)

Old Stratford Bridge – south side

tail of the rounded cutwaters

39

The 1835 bridge we see today is constructed of red dressed stone, has three stone arches supported by two stone piers with rounded cutwaters, and the height was raised by a further 1.2 metres to prevent any flooding. The road that crossed the bridge was just enough for two horse-drawn carriages to pass one another at the same time between stone parapets. There is a decorative cornice between the parapets and the bridge substructure. The causeway running ¼ mile to the east is made of smaller local stone and contains three brick lined flood arches to allow adequate drainage during times of high water. The bridge then, as now, was surrounded by meadows and some beautiful scenery. It was carried across the river from the higher ground at Old Stratford and continued with a raised causeway across the flood meadow on the Stony Stratford side.

The bridge still stands 150 years later, carrying the huge increase in today's traffic. The only difference is the absence of the Toll House, which originally stood at the southern (Stony Stratford) end. This was removed in 1857 despite a petition by 69 local inhabitants, amongst whom are to be found most of the 'solid citizens' of the time, who expressed concern that this would lead to damage to the bridge and danger to people using it.

The three flood arches to the south within the causeway

Detail of the soffits below the parapets

Old Stratford Bridge as it looked in 1989 (© Transport for Buckinghamshire)

The road carriageway of the Old Stratford bridge as it appeared in 1914 (© Transport for Buckinghamshire)

Old Stratford bridge around the turn of the century looking south into Stony Stratford (© Transport for Buckinghamshire)

Old Stratford Bridge, early 1900s looking south across the River Great Ouse (© Transport for Buckinghamshire)

Bridge Name:	London Road Bridge
Other Names:	Buckingham Bridge, Long Bridge and the A413 Bridge
Location:	Central Buckingham
National Grid Reference:	SP 69697 33825
Crosses:	River Great Ouse
Span:	17 metres
Bridge Type:	Three Arches
Materials:	Ashlar Stone
Traffic:	Pedestrian & Road
Opened:	1805
Managed by:	Transport for Buckinghamshire
Historic England Designation:	Grade II Listed

Information: Buckingham's London Road Bridge has been the main route into town for more than 200 years. Samuel Pepys, who visited Buckingham on 8 June 1668, had this to say about it:

A good old town. Here I to see the Church, which is very good, and a school in it: did give the sexton's boy 1s. A fair bridge here with many arches.

(Buckingham Society, 2016)

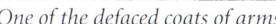

The Buckingham swan formed of coade stone. Also visible is one of the defaced coats of arms

The current bridge was built in 1805 at the expense of the Marquess of Buckingham to create a pleasant route for the future King George IV on his way to a party at the Marquess's home at Stowe. It is constructed of dressed stone with parapets on both sides. There are three arches with two centre piers with stone voussoirs and clearly offset keystones. The piers have cutwaters and (now defaced) florets. In the centre of the south side of the parapet is the Buckingham swan of Coade stone. The abutments and wing walls are made of the same material as the bridge and the north wing wall also has a defaced coat of arms. There are circular 'florets' on the north side hidden behind the modern footbridge, which sits only 0.6 metres away.

When built in 1805 the London Road Bridge became the main town entry point and replaced a narrow stone bridge known as the Sheriff's Bridge that had been built in the 16th century, next to the Woolpack Pub. This was the old coach-road approach from London before the present high road was cut and served as the main Ouse River crossing for many years but was in a bad state

One of the defaced coats of arms

The restored Marquess of Buckingham's coat of arms made from Coade stone

of dereliction and likely to fall. This itself has been replaced by an iron lattice footbridge adjacent to the road bridge. *(British History Online – Buckingham, 2004; Buckingham Society, 2016)*

A postcard from 1909 showing what they then called 'The Long Bridge'
(© Buckinghamshire Archives, 2020)

Buckingham's London Road Bridge looking north

The south elevation – London Road Bridge

In March of 2016, the Marquess's coat of arms in the centre of the north side of the bridge, which had been seriously vandalised several years earlier, was restored in Coade stone by the Buckingham Society.

The bridge is a Grade II listed structure.

14 Old Farm Bridge

Bridge Name:	**Old Farm Bridge**
Location:	**Within Bourton Park, southeast corner of Buckingham**
National Grid Reference:	**SP 70866 33498**
Crosses:	**River Great Ouse**
Span:	**3.5 metres**
Bridge Type:	**Single Arch**
Materials:	**Brick with Some Stone**
Traffic:	**Pedestrians & Cattle**
Opened:	**Unknown**
Managed by:	**Bourton Park**
Historic England Designation:	**Unlisted**

Old Farm Bridge – Bourton Park, Buckingham

Old Farm Bridge, slightly closer view

Information: This very old bridge lies within Bourton Park at the southeast corner of Buckingham, just off the A413. Very little information can be found about this bridge, but its picturesque nature will have made it familiar to many photographers. The River Great Ouse passes through this portion of Bourton Park within a walled aqueduct. To access it start within the Park's car park and walk to the right (east) along the river. Pass under the new A413 bridge (low headroom) and the old bridge is less than 100 metres ahead. The brick arch of 3.5 metres span now sits on the concrete aqueduct walls and it is apparent that some care was taken to preserve the structure when it would have been much easier to demolish it. There are no parapets but there are dilapidated pipe and concrete guardrails that are not the same age as the bridge. The bridge itself is overgrown and the worse for wear. Without some sympathetic maintenance and attention, we won't have this pleasant little structure for much longer.

Bridge Name:	Lord's Bridge
Location:	Hunter Street in central Buckingham near the University of Buckingham. Just near the overhead railway bridge.
National Grid Reference:	SP 69316 33435
Crosses:	River Great Ouse
Span:	31 metres
Bridge Type:	Double Arch
Materials:	Red Brick
Traffic:	Pedestrian & Road
Opened:	c.1860
Managed by:	Transport for Buckinghamshire
Historic England Designation:	Unlisted

Information: A lovely red brick bridge that sits today on the campus of the University of Buckingham. Lord's Bridge carries Hunter Street over the River Great Ouse. It has two arches constructed from five layers of brick sitting on a central pier. The pier has a rounded cutwater on the west side. Starting at the string course, the parapets are slightly extended outward from the substructure and are also made of brick with dressed stone coping.

Buckingham's Lord's Bridge, west side

The 'Flosh' weir just below Lord's Bridge in Buckingham

Lord's Bridge and the weir as they looked in 1990
(© Buckinghamshire Archives, 2020)

The roadway is 7.2 metres wide with pedestrian sidewalks. Publicly accessible parkland surrounds the bridge and the east side has a weir named the 'Flosh' that creates a popular wading pool.

Stone capping on the parapets of Lord's Bridge in Buckingham

The central pier showing the rounded cutwater and the five brick layered arches – Lord's Bridge

16 Cosgrove Iron Trunk Aqueduct

Bridge Name:	**Cosgrove Iron Trunk Aqueduct**
Location:	**Cosgrove. It can be accessed on foot (along the canal towpath) from Old Wolverton Road, between Stony Stratford and Wolverton.**
National Grid Reference:	**SP 80052 41825**
Crosses:	**River Great Ouse**
Span:	**31 metres**
Bridge Type:	**Arch**
Materials:	**Cast Iron & Brick**
Traffic:	**Aqueduct & Pedestrian**
Opened:	**1881**
Managed by:	**Milton Keynes Council**
Historic England Designation:	**Grade II Listed**

Cosgrove's Iron Trunk Aqueduct viewed from the south

View of the aqueduct from the pedestrian footbridge

Date stones in the southwest parapet column

Transport Trust sign

Wrought iron 'retaining' straps on the column's capstones

Information: Cosgrove Iron Trunk Aqueduct is a navigable cast iron trough that carries the Grand Union Canal over the River Great Ouse, on the border between Buckinghamshire and Northamptonshire at the northwest margin of Milton Keynes. There is a footpath (originally a towpath) set some 12 metres above the river's surface. It is a dazzling (and in some cases, dizzying) route with views as far as the eye can see over the Buckinghamshire and Northamptonshire countryside.

The challenge facing the Grand Union Canal Company at this site was that their canal was 12.1 metres higher than the River Great Ouse valley it had to cross. Initially, flights of locks, four at the southeast and five at the northwest, were used to allow canal boats to descend to cross the river on the level and then raise them again to the level of the canal. This was a slow, laborious, and resultantly expensive process. Then William Jessop, the canal company's engineer, designed a three-arch brick viaduct so that the canal could cross at the higher level, reducing the water loss and long delay in locking down to river

level and then back up again. His structure was opened on 26 August 1805, but a section of the canal embankment collapsed only five months later in January 1806; this was repaired, but then two years later the aqueduct structure itself collapsed in February 1808, severing the canal. The present structure was built in 1811, to replace the previous brick structure that had failed.

When this present-day 'wide canal' structure was erected, it was known as the 'Iron Trunk'. The structure has two cast iron troughs that span the river below with a single central masonry pier in the centre. The abutments were constructed in stone but have been refaced in brick during the 20th century. Each trough is 4.6 metres wide, 1.98 metres deep, with a total length of 31 metres. The canal surface is about 12 metres above the surface of the river. There are large approach earthworks about 11 metres high above the valley floor and 46 metres wide, with a total length of 800 metres. (*Engineering Timelines, 1999*)

Entrance to the cattle creep tunnel dated 1919

The new cast iron troughs were cast at the Ketley Iron Works at Coalbrookdale; the company had already been successfully involved with the Longdon-on-Tern Aqueduct. The castings were transported to Cosgrove by the canal itself and then assembled and erected on site. The new structure was completed in January 1811.

To support the substantial additional weight of iron, water, and traffic of a wide canal, the floor sections of the troughs were designed to be arched, providing additional strength. In addition, there are arch ribs built within the trough side plates. The towpath is cantilevered from one side and supported by diagonal struts. However, at Cosgrove the towpath is cantilevered outwards over the river, rather than inwards over the canal.

There is an interesting cattle creep (tunnel) through the embankment on the south side of the main structure with a date stone of 1919 above the southern entrance. Cattle creeps are small field-to-field access openings for farm animals and pedestrians.

In 2011, the Cosgrove Iron Trunk Aqueduct won the final round of 'The Big Lottery Fund: The People's Millions' – with a prize of £60,000 in funding. The money was used to celebrate the aqueduct's 200th birthday by cleaning and repainting the cast iron structure in its original colours.

RIVER OUZEL

The River Ouzel (also known as the Lovat) rises from chalk springs near Dagnall in the Chilterns. It flows northwards for 20 miles through Leighton-Linslade, cutting a gap through the wooded sandstone hills of the Greensand Ridge before passing through the east side of Milton Keynes and joining the River Great Ouse at Newport Pagnell. For some of its length it runs beside the Grand Union Canal and at Twelve Arches in Linslade the overflow from the canal runs down into the river across the flood plain. The Clipstone and Broughton Brooks are two of its main tributaries.

One theory for the origin of the name 'Ouzel' is a historic association with the dipper, a thrush-sized bird now largely associated with upland rivers and streams, but which is featured on the town crest of Leighton-Linslade.
(The Upper & Bedford Ouse Catchment Partnership, 2016)

17 Tickford Iron Bridge – FEATURE BRIDGE

Bridge Name:	Tickford Iron Bridge
Location:	North of Milton Keynes, where the B526 crosses the River Ouzel in central Newport Pagnell.
National Grid Reference:	SP 87775 43850
Crosses:	River Ouzel
Span:	18 metres
Bridge Type:	Arch
Materials:	Cast Iron with Sandstone Abutments
Traffic:	Pedestrian & Road
Opened:	1810
Managed by:	Milton Keynes Council
Historic England Designation:	Grade I Listed

Information: Tickford Iron Bridge carries Tickford Street over the River Ouzel (or Lovat) in Newport Pagnell. It was built in 1810 as a toll bridge and is one of the last (21 still remaining) cast iron bridges in Britain that still carries modern road traffic, though it has had some strengthening. This well-known structure is believed to be the oldest iron bridge in the world that is still in constant use.

It was cast in iron by Walkers Iron Works of Rotherham, in Yorkshire. There is a plaque near the footbridge at the side that gives details of its history and construction. The bridge was jointly designed by Thomas Wilson (1751–1820) and Henry Provis (1760–1830) and followed Wilson's patented method for the construction of such arches.

Cast iron sign on bridge

The abutments are carried down to bedrock at the riverbanks and are built of local sandstone. They have piers at either end and coping, with bands at the base of the parapet.

The superstructure is formed by six equally spaced cast iron ribs connected by cast iron transverse diaphragms. Each rib is composed of 11 segments, with mortice and tenon joints keyed together to form an arch.

The spandrel panels consist of iron rings diminishing in size towards the crown of the span. The original deck plates are of iron and are placed over the upper members of the spandrel panels and located by continuous longitudinal lugs. Iron railings complete the design.

Following the fracture of one of the deck plates in 1900, the three inner sections were strengthened by the addition of wrought iron arch plates, bolted through the iron deck plates

Tickford Iron Bridge lamps needing repair

Tickford Iron Bridge

The Tickford Iron Bridge 1967 (© Buckinghamshire Archives, 2020)

close to the ribs. In 1976, a reinforced concrete slab was laid over the deck on a 20 mm cushion of plastic foam. *(Harris, 1968)*

In 2017 the two original brass lamps in the centre of the bridge were removed and repaired.

According to 'British Listed Buildings', Tickford Iron Bridge is a foremost work of early cast iron engineering, still surviving in near original condition and taking modern traffic. It is a monument of national importance in the history of civil engineering and the use of cast iron. Tickford Iron Bridge is Grade I listed by Historic England.

Tickford Iron Bridge 1820 (Artist Unknown) (© Newport Pagnell Historical Society, 2020)

Tickford Iron Bridge 1912 (© Buckinghamshire Archives, 2020)

This drawing of 1820 shows the premises on the opposite side of the famous Iron Bridge only a few years after its construction. This sketch is looking downstream towards Rivers Meet (where the Great Ouse meets the River Lovat) and shows Bridge House. The premises were originally a fellmongers yard and tannery in the occupation of the Yates family. *(Newport Pagnell Historical Society – History, 2020)*

Tickford Iron Bridge
(© Buckinghamshire Archives, 2020)

Detail of Tickford Iron Bridge lamps

STOWE LANDSCAPE GARDENS

The estate of Stowe, just north of Buckingham in the attractive agricultural area of Aylesbury Vale, was owned from the late 16th century by the Temple family, who made their fortune from sheep farming in the Vale. They built a grand house surrounded by extensive gardens beginning around 1716, chiefly while Sir Richard Temple, the 1st Viscount Cobham, owned the property. Much of his design can still be seen today. The development of the grounds continued under his son George Temple-Grenville up until 1846 when the family fortune ran out. The estate remained in ownership of his nephew Richard Grenville, Earl Temple, until 1921.

In March 1741, Lancelot 'Capability' Brown was appointed by Grenville to be head gardener at Stowe following William Kent. He worked with the architect of the grounds, James Gibbs, until 1749. Brown oversaw paying not just the under-gardeners at Stowe but also the carpenters, masons, and craftsmen working on the house and other buildings. The estate accounts show that in 1742–1743 Brown was ordering stone from three different quarries and overseeing masons working in the library and chapel at the house, and on the Palladian Bridge. The gardens now belong to the National Trust, but Stowe House (not National Trust) is occupied by Stowe School.

Bridge Name: Palladian Bridge

Location: Within the grounds of Stowe Landscape Gardens, northwest of Buckingham.

National Grid Reference: SP 68002 37192

Crosses: Private Lake

Span: 15 metres

Bridge Type: Three Arches plus two Decorative Arches

Materials: Masonry Ashlar Stone

Traffic: Horse-Drawn Carriages & Pedestrians

Opened: 1738

Managed by: National Trust

Historic England Designation: Grade I Listed

Information: This bridge sits within the grounds of the well-known Stowe Landscape Gardens near Buckingham, which are managed by the National Trust. Palladian Bridge is a copy of the bridge at Wilton House near Salisbury in Wiltshire. Three more copies and variations were erected, at Prior Park in Bath, Hagley Hall in Worcestershire, and Amesbury in Wiltshire. Empress Catherine the Great even commissioned another copy, known as the Marble Bridge, to be set up at the landscape park of Tsarskoye Selo in St Petersburg, Russia.

Palladian Bridge – Stowe (© National Trust Images)

An English Landscape Architect

Lancelot 'Capability' Brown (1716–1783)

Lancelot Brown changed the face of 18th-century England, designing country estates and mansions, moving hills, diverting rivers, and creating lakes. He is mentioned many times in this book not only for his own work but also for his influence on those landscape artists that came after him. Lancelot Brown was a leader in the development of the 'natural', 'English', or 'serpentine' style of gardening.

Lancelot Brown was born in Northumberland and served as an apprentice with Sir William Lorraine. A move to Buckinghamshire in 1739 led to his employment by Lord Cobham at Stowe in 1741, where his job as head gardener was to last 10 years.

It was Stowe that gave Brown the opportunity to work with William Kent and see great works carried out there under the overall direction of Kent.

In 1751, Brown was an independent landscape gardener, although he described himself as a 'place-maker' rather than a landscape gardener, and quickly became very fashionable and in great demand.

Lancelot Brown was known as 'Capability' because of his fondness for speaking of a country estate having a great 'capability' for improvement. He rejected the very formal geometric French style of gardening (see G. London & H. Wise later in this text), a perfect example being at Versailles, and concentrated on echoing the natural undulations of the English landscape.

Characteristics of his work included grass meadows in front of the mansion, serpentine lakes, follies, encircling carriage drives with bridges, belts, and circular clumps of trees. Bridges or cascades were often used to connect the 'natural' lakes and a great many formal gardens were destroyed on Brown's say-so, to be replaced with landscapes, which did lead to criticism later.

Lancelot 'Capability' Brown's career of 32 years saw his style hardly change, with the serpentine shapes becoming his hallmark. His popularity peaked at the time of his death, but then fell into decline, reaching its lowest point in the 1880s. By 1980 however, after a gradual recovery, Lancelot Brown was recognised as a genius of English garden design.

Lancelot 'Capability' Brown's sympathetic method of working meant that of the 200 plus parks he designed, a surprising number remain in good condition. Often, they have adapted well to modern-day use as public parks, farms, golf courses, and schools. Some of the estates he designed are included in this book.

Image public domain

The southern entrance to the pavilion
(© National Trust Images)

The four central columns of the pavilion. The two columns on each end are half columns.
(© National Trust Images)

The ornate plaster ceiling of the pavilion
(© National Trust Images)

View through the pavilion
(© National Trust Images)

The main difference between these bridges and Stowe's is that the Stowe version was designed to be used by horse-drawn carriages, so is set lower with shallow ramps instead of steps on the approaches. It was completed in 1738 under the direction of James Gibbs.

The substructure of this bridge has five arches, the central one is wide and segmental with a carved head on the keystone. The two flanking semi-circular arches are smaller and with identical carved heads of a young woman on their keystones. At each end of the bridge there are semi-circular faux arches (alcoves) that do not carry water flow. There is a balustraded parapet that runs the length of the bridge, abutments, and wing walls.

The central arch supports an open pavilion above, acting as a covering for the carriageway. This pavilion consists of colonnades of four full and two half columns of unfluted Roman Ionic order. Above the flanking arches there are pavilions with arches on all four sides. These have engaged columns on their flanks and ends of the same order as the colonnade, which in turn support pediments. The roof is of slate, with an elaborate plaster ceiling.

Palladian Bridge originally crossed a stream that emptied from the Octagon Lake, and when the lake was enlarged and deepened, and made more natural in shape in 1752, this part of the stream became a branch of the lake. *(Geni, 1999)*

The three arches with carved keystones (© National Trust Images)

One of the two identical carved keystones (© National Trust Images

The Palladian Bridge southern approach (© National Trust Images)

Architect and Landscape Designer – Palladian Bridge

James Gibbs
(1682–1754)

At the age of 20, James Gibbs travelled to Rome to begin his training as a Catholic priest. However, within a year of being surrounded by grand Italian design, he had thought better of it, deciding instead to study architecture. Completing his training as an architect in Rome, James Gibbs brought back to Britain a full understanding of Italian Baroque design. Early in his professional career, however, he was forced to adapt his ideas for buildings to suit the prevalent taste powerfully advocated by Lord Burlington who held a more austere view of Classical architecture informed by the work of Inigo Jones and the Vincentian architect Palladio (Palladian).

Gibbs is most well-known for his work at Stowe where he had two separate engagements. At Stowe he created the final and greatest architectural elements of the estate, most noteworthy being the Palladian Bridge. He started work on the bridge, crossing the far end of Octagon Lake, linking the two sides of the Hawkwell Fields in 1759. The bridge design is based on a similar one built at Wilton House in Wiltshire a year earlier. Gibbs observed the construction at Wilton but altered the design slightly. The bridge deck used at Stowe was lower and wider and without steps, allowing for carriages to cross on tours of the gardens. The bridge would later form part of the Path of Liberty, the longest walk through the gardens, which also takes in most of the other garden additions from Gibbs.

Through his publication *A Book of Architecture, Containing Designs of Buildings and Ornaments*, London 1728, Gibbs also made himself one of the most influential British architects of the 18th century. This helped formulate the language of what is generically termed 'Georgian' design.

The Palladian Bridge in autumn at Stowe (© National Trust Images/Hugh Mothersole)

Bridge Name:	Oxford Bridge
Location:	Within the grounds of Stowe School, northwest of Buckingham.
National Grid Reference:	SP 66700 36797
Crosses:	Private Lake (River Dad)
Span:	11 metres
Bridge Type:	Three Stone Arches
Materials:	Rustic Stone
Traffic:	Pedestrian & Road
Opened:	1761
Managed by:	National Trust
Historic England Designation:	Grade I Listed

Information: Like the Palladian Bridge, the Oxford Bridge sits within grounds that were formerly the estate of Viscount Cobham and have been managed by the National Trust since 1989. The Viscount Cobham estate remained in ownership of his nephew Richard Grenville, Earl Temple until 1921, when the house became Stowe School.

Oxford Bridge was built in 1761 to cross the river Dad after this had been dammed to form what was renamed the Oxford Water. Its hump-backed form was probably designed by Richard Grenville, 2nd Earl Temple (1711–1779) as part of the western entrance avenue to the estate. It is built of rustic hewn stone, which rises to form the slightly inset parapets. There are three arches, the central one being slightly wider and higher than the flanking ones. The three arches sit on two stone piers, which, although this is a lake, have small cutwaters on both sides. The abutments and wing walls help traverse the width of the lake.

There are four decorative stone urns with carved faces placed at the ends of the parapets, and four more placed above the two piers. These were probably brought here from the former Temple of Sleep, since demolished. *(Geni, 1999)* This is a Grade I listed bridge.

The northern side of Oxford Bridge (© National Trust Images)

Oxford Bridge 1979 (© Buckinghamshire Archives, 2020)

Detail of the stone arches and piers (© National Trust Images)

Visitors on the Oxford Bridge on a frosty day at Stowe Landscape Gardens, Buckinghamshire (© National Trust Images/Rod Edwards)

One of four identical urns (© National Trust Images)

Oxford Bridge – Stowe (© National Trust Images)

Oxford Bridge – Stowe (© National Trust Images)

The western entrance to the Cobham estate, now Stowe School (© National Trust Images)

The carriageway and northern parapet with urns (© National Trust Images)

53

Bridge Name: **Shell Bridge**

Location: **Within the grounds of Stowe Landscape Gardens, northwest of Buckingham.**

National Grid Reference: **SP 67739 37341**

Crosses: **Private Lake (River Dad)**

Span: **9.5 metres**

Bridge Type: **Five Stone Arches**

Materials: **Rustic Stone**

Traffic: **Pedestrian**

Opened: **1740**

Managed by: **National Trust**

Historic England Designation: **Grade I Listed**

Shell Bridge and the Temple of British Worthies (© National Trust Images)

Information: Shell Bridge is actually a dam in Stowe's Elysian Fields, with a low-profile bridge helping disguise its functional nature. Like the previous two bridges it sits within the grounds of Stowe Landscape Gardens. It was built around 1740 to designs by William Kent.

It is rustic in style, made of stone with rendering encrusted with shells. The bridge has five low arches. The larger central arch protrudes from the bridge face slightly and has a small

Shell Bridge – Stowe (© National Trust Images)

Shell Bridge – Stowe (© National Trust Images)

individual pediment as a parapet. There are two smaller arches to each side that sit on stone piers. Each pier has its own pediment, which along with upturned field stone, form a parapet along the length of the bridge. The abutments sit slightly into the lake. There is a small monument to Captain Cook standing on the bridge with a portrait medallion dated 1778.

RIVER THAME

The River Thame (pronounced 'Tame') rises in the northern Chilterns as a combination of several streams to the north of Aylesbury. It meanders for about 40 miles through Buckinghamshire and Oxfordshire, past the town named after it until it meets the Thames at Dorchester-on-Thames in Oxfordshire. At most times it is a fairly insubstantial river, especially in its more northerly reaches, but after heavy rains it can become far bigger, such as after the prolonged series of downpours experienced from December 2013 to February 2014 when it burst its banks and flooded the neighbouring land, mostly farmland, in its wide shallow valley.

21 Holman's Bridge

Bridge Name:	Holman's Bridge
Location:	A413 in Aylesbury just south of Watermead Road and north of Oliffe Close near the Horse & Jockey Pub.
National Grid Reference:	SP 81774 15270
Crosses:	River Thame
Span:	9 metres
Bridge Type:	Three Arches
Materials:	Brick
Traffic:	Pedestrian & Road
Opened:	1842
Managed by:	Transport for Buckinghamshire
Historic England Designation:	Unlisted

Information: Holman's Bridge is a brick-built bridge to the north of Aylesbury where the A413 road crosses the River Thame. It is constructed of three arches with a span of 2.5 metres each. The brick rises to form low parapets with stone coping.

Aylesbury's first Charter of Incorporation in 1554 marked Holman's Bridge as the northernmost boundary of the town. On 1 November 1642 the battle of

The three arches of Holman's Bridge

Aylesbury took place here between the Royalists under the command of Prince Rupert and the Parliamentarians, under the command of Sir William Balfore. Although vastly outnumbered, Balfore's troops defeated the Royalists, and the engagement became known as the Battle of Aylesbury.

In 1818 a large quantity of human bones were discovered buried together near to Holman's Bridge. The condition, artefacts, and location identified them as being the remains of the men who had fallen in the battle. The bones were collected and deposited in a tomb in the churchyard in the village of

Holman's Bridge on the A413

The monument to the soldiers in the churchyard at Hardwick

Hardwick, at the request of the late Right Hon. Lord Nugent and the following inscription was engraved on the tablet:

Within are deposited the bones of 247 Persons who were discovered A.D. 1818, buried in a field adjoining to Holman's Bridge, near Aylesbury. From the History and appearances of the place where they were found, they were considered to be the bones of those officers and men who perished in an engagement fought A.D. 1642, between the troops of K. Charles I., under the command of Prince Rupert, and the Garrison who held Aylesbury for the Parliament. Enemies from their attachment to opposite leaders and to opposite Standards, in the sanguinary conflicts of that Civil War, they were together victims to its fury. United in one common slaughter, they were buried in one common grave, close to the spot where they had lately stood in arms against each other. After the lapse of more than a century and a half their bones were collected and deposited still in consecrated ground. May the memory of brave men be respected, and may our country never again be compelled to take part in a conflict such as that which this tablet records.

In 2006 work began to add a wooden pedestrian bridge alongside the existing bridge to provide pedestrian access to the new Weedon Hill housing estate. The housing development is controversially built on the tract of land where the battle took place. The ensuing arguments about the land use led to debate as to whether Oliver Cromwell was actually at the battle and indeed whether the battle itself really took place there at all. Documents in the Buckinghamshire County Archives describe the incident as 'probably a skirmish rather than a battle' and that it is 'significant that Prince Rupert's journal records him at Abingdon on the date of the supposed battle, and makes no mention of any severe setback to Royalist forces in Bucks'.

What is beyond doubt is that Oliver Cromwell came back to Aylesbury after the Battle of Worcester, and that he stayed in the King's Head. On arrival at the King's Head he received a Vote of Thanks letter from Parliament. The National Trust, who now own the King's Head in Aylesbury, has several artefacts dating from this Civil War incident. These include anti-monarchy tokens given out by the then landlord, which could be exchanged for ale, and the chair said to have been used by Cromwell during his stay. We will probably never know for certain whether Oliver Cromwell was at the battle of Holman's Bridge, but his association with the town has echoes that reverberate even today. *(Peters, 2009)*

Today's bridge is an unimpressive structure which thousands pass over every day without noticing.

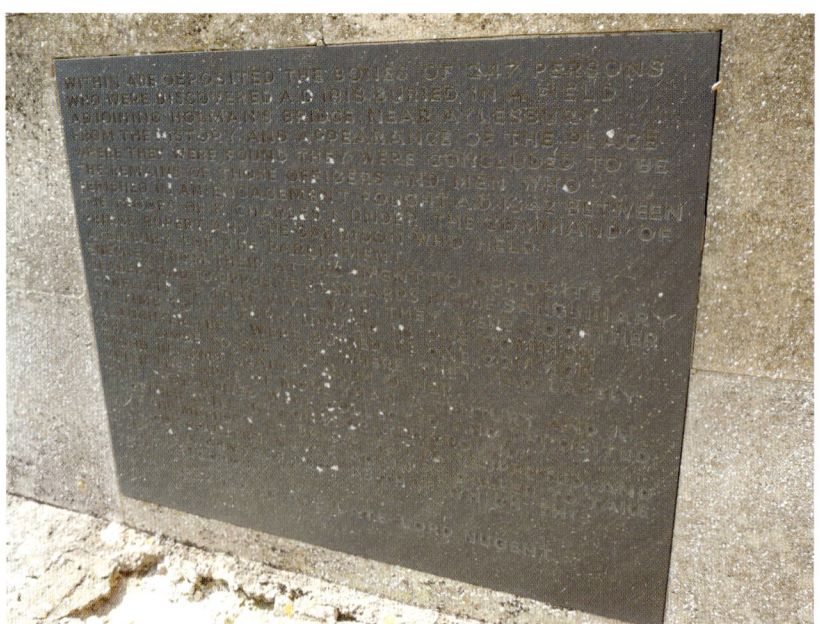

The engraving on the monument

Architect –
Eythrope Park Bridge

Isaac Ware is known primarily for his writings on architecture and left a legacy of designs that influence architecture today. He was born in London, the son of a shoemaker, in 1704. Ware was reputed to have started as a chimney-sweep boy and to have been spotted in the street drawing the elevation of the Banqueting House by a gentleman, possibly Lord Burlington, who took him on as his charge. This story is, however, undocumented. Whoever his benefactor was he was given a good education and sent to Italy to study Renaissance architecture where he was greatly influenced by the Palladian style. On his return he formed connections with Lord Burlington's influential circle and may have been patronised by Burlington himself.

Isaac Ware
(1704–1766)

Ware wrote several important works on architecture including *Designs of Inigo Jones and Others*, which contained designs by Burlington. At the same time Ware produced his most important work *The Complete Body of Architecture*. This was commissioned by the publishers Osborne & Shipton as a companion to similar treatises by Ware on animal husbandry and gardening. The *Complete Body*, as it became known, contained a statement on Georgian architectural theory which balanced adherence to Palladio's principles with the exercise of imagination and judgement.

Ware's output as an architect was very limited. Apart from a few country houses, including one in Scotland, Amisfield (1755 – demolished 1928), his major works were Lord Chesterfield's town house in South Audley Street, London (1748–1749), Clifton Hill House in Bristol (1746–1750 – his most complete surviving building), and the former Town Hall in Oxford (1751–1752). He was also involved in speculative building in London.

Most of Ware's designs were Palladian in character. However, he also designed in a lavish Rococo style – for example, the interiors of Chesterfield House. As a prominent member of St Martin's Academy, he rubbed shoulders with Rococo artists like Hogarth, Roubiliac, and Francis Hayward. Ware looked beyond straightforward Palladian theory, and this is supported by what he wrote in *The Complete Body of Architecture*.

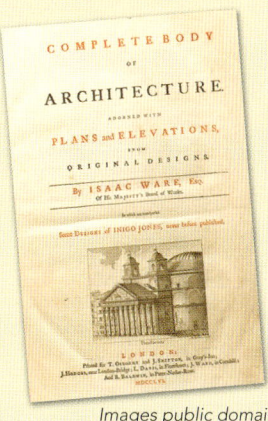

Images public domain

Bridge Name:	Eythrope Park Bridge
Location:	2.5 miles west of Aylesbury. Turn north off the A418 in Stone onto Eythrope Road and follow to dead end and walk. Within grounds of Eythrope Park.
National Grid Reference:	SP 76772 13492
Crosses:	River Thame
Span:	9 metres
Bridge Type:	Single Arch
Materials:	Masonry and Dressed Stone
Traffic:	Pedestrian
Opened:	1741
Managed by:	National Trust
Historic England Designation:	Grade II Listed

Information: Eythrope Park is a garden estate now looked after by the National Trust along with the adjoining Waddesdon Manor estate. Eythrope existed as a separate estate prior to its purchase in 1875 by Alice de Rothschild (1847–1922) a year after her brother acquired the adjacent Waddesdon Manor estate to the north. She developed 74 acres of highly ornamental and innovative gardens, with a large kitchen garden, surrounded by the existing park.

Eythrope Park Bridge   view of the west side

Eythrope Park Bridge

Along the riverbank

Eythrope House was demolished in 1811 and replaced in the late 19th century with a day pavilion. The dressed-stone bridge stands 200 metres southwest of where the house once stood and spans the River Thame with a wide single segmental arch with rusticated voussoirs and a carved stone mask head on the keystone. Originally built with parapets for Sir William Stanhope, Isaac Ware designed the bridge but it was built in 1738 by Banister Watts, a High Wycombe mason whose account-book shows that he 'began the bridge work at Eythrope' on 5 June 1738. Today's bridge is unhappily shorn of its parapets but with more recent iron railings, which have been added later. The stone for the construction of the bridge is said to have come from Eythrope Chapel, which Stanhope had taken down in 1738. There is a small waterfall on the east side of the bridge. (*British History Online – Waddesdon, 2001*)

The roadway and railings – Eythrope Park

Bridge Name:	Wotton Underwood Bridge
Other Names:	Five-Arch Bridge
Location:	Wotton House near Brill. Wotton House lies at the northwest corner of Wotton Underwood village, in the Vale of Aylesbury, 6 miles north of Thame and 8 miles west of Aylesbury.
National Grid Reference:	SP 67796 15930
Crosses:	River Thame
Span:	9 metres
Bridge Type:	Five Arches
Materials:	Masonry and Stone
Traffic:	Pedestrian
Opened:	1757–1760
Managed by:	Privately
Historic England Designation:	Grade II Listed

Wotton's five-arch bridge and dam

Information: This bridge is within the Wotton Underwood House Estate grounds, a mid-18th-century park, landscape, and woodland originally designed by Richard Grenville, father of George Grenville who served as British Prime Minister from 1763 to 1765. It was George Grenville's imposition of new duties on the American colonies that helped provoke the outbreak of the American War of Independence.

Wotton House was surrounded by formal garden enclosures, and Richard Grenville seems to have laid out the landscape with the help of George London and Henry Wise. His son, George Grenville, subsequently redesigned the gardens with a contemporary layout. This was consequently developed into an extensive 'naturalised' park by Lancelot 'Capability' Brown. William Pitt, who later became the first Earl of Chatham, is credited with significant input regarding the design. Lancelot 'Capability' Brown worked at Wotton briefly in 1739–1740, and it is believed that he returned to Wotton over several years in the 1750s, during which time he landscaped the park. The landscape remained largely undisturbed after the mid-18th century. The grounds are currently undergoing restoration and are related to the Stowe Gardens to the extent that both belonged to the Grenville family when Brown laid out the Wotton grounds.

There is a large, irregularly shaped lake that lies at the north end of the

Wotton House from the lake

grounds and contains several islands, the largest of which is Grotto Island towards the southern end.

The lake narrows to run south as a curving canal, crossed by a wooden Palladian bridge (late 20th-century reconstruction), and then on to the stone-faced 'five-arch bridge' at its southern end. This bridge disguises a low dam with an overflow from the canal via the central arch. It was probably built around 1758–1760, to a design by Sanderson Miller, who is known to have designed a bridge for George Grenville in 1758. The structure is modelled on William Kent's Shell Bridge at Stowe.

The bridge is constructed of coursed rubble stone with ashlar dressings. The piers have ashlar plinths, quoins, and pedimented copings. There is an ashlar band along the top of the bridge but without parapets. The bridge has five segmental arches, with the central arch being slightly projected with a pediment to the west side. This arch has the overflow channel to the lower lake. The bridge has outwardly splayed wing walls at each end. The east side of the bridge is damaged.

Below the bridge on the east side, the dam has

20th-century Palladian bridge

Crossing the Palladian bridge

View from the Palladian bridge

matching stone facing with blind arches flanking a central overflow arch. The water then flows into The Warrells, a symmetrical lake naturalised by Brown during the 1750s. At its west end it corners into narrow channels, with the 'five-arch bridge' terminating the southwest corner and the northwest corner terminated by China Island.

(Historic England – Five Arch Bridge, 2016)

Landscape Artists and Plantsmen – Wotton Underwood

Henry Wise
(1653–1738)

George London (c. 1640–1714) and Henry Wise must vie with Gertrude Jekyll and Edwin Lutyens for the title of Britain's greatest gardening double act. As sole partners at the renowned Brompton Park Nursery from 1689 until London's death in 1714, they enjoyed a near monopoly on large-scale landscape design, also supplying thousands of trees to landowners for avenue planting.

London and Wise specialised in an English version of the formal Baroque gardens associated with the Catholic courts of continental Europe, of which Versailles was the pre-eminent example. These were gardens in which magnificent flat parterres spread out below one or two façades of the palace or house, defined by box hedges in patterns derived from textile designs and enlivened with coloured gravels, white or painted statuary, extravagant fountains and colourful annual flowers.

These were the predecessors and antithesis of 'Capability' Brown's ideals of natural simplicity. In fact, many of London and Wise's formal gardens were wiped out in the following decades by Brown or his followers. None of London and Wise's designs survive today, although the restored Privy Garden at Hampton Court Palace (probably designed by Daniel Marot) provides a good idea of the style.

Wise is described as the business brain of the outfit, and it does appear he oversaw nursery operations, although, as a designer, he was responsible for the great formal garden made at Blenheim for the Duke of Marlborough. London was the star, however. He worked all over England (as Brown was to do) creating formal gardens at great houses such as Longleat, Chatsworth, and Burghley, and his role as an early exponent of the naturalistic style has only recently begun to be appreciated.

Palladian bridge in the landscape

Bridge Name:	Bridgeway Bridge
Other Names:	Cuddington Bridge
Location:	On Aylesbury Road after Chearsley and before Cuddington near Aylesbury.
National Grid Reference:	SP 72871 11327
Crosses:	River Thame
Span:	12 metres
Bridge Type:	Three Arches of Triple Brick
Materials:	Red and Black Brick
Traffic:	Pedestrian & Road
Opened:	Mid-19th century
Managed by:	Transport for Buckinghamshire
Historic England Designation:	Unlisted

Information: This bridge goes unnoticed as you drive from Chearsley to Cuddington. The road, at the point of crossing the River Thame, is called Bridgeway Road, associated with Bridgeway House, and lies in a picturesque valley. This bridge is unusual in that it utilises two colours of brick with the lighter on the substructure and the parapets are of a darker black and

Bridgeway Bridge at Cuddington

brown brick. There is no record of the substructure and the parapets being constructed separately. There are three 2 metre arches and there are brick cutwaters on the north side. The south side of the riverbed has been recently cleared by the Council.

This bridge has always been historically important to the Nether Winchendon Estate and once formed a part of the main route from Aylesbury.

On the riverbank

The cutwaters on the north side – Bridgeway Bridge at Cuddington

View across the valley with a few of the local residents

Bridge Name:	Hartwell House Bridge
Location:	Between Stone and Aylesbury on the A418. The house is about 2 miles northwest of the village of Stone about 3 miles from the centre of Aylesbury.
National Grid Reference:	SP 79672 12608
Crosses:	Private Water – River Thame tributary possibly Bear Brook
Span:	8 metres
Bridge Type:	Arch
Materials:	Stone
Traffic:	Pedestrian & Road
Opened:	Early 19th century
Managed by:	National Trust
Historic England Designation:	Grade II Listed

Hartwell House Bridge (© National Trust Images)

Information: Hartwell House is a country house in the village of Hartwell between Stone and Aylesbury on the A418. The house is part of the Hartwell Estate owned by the Ernest Cook Trust, and since 2008 has been leased to the National Trust. It is a Grade I listed building and is currently a part of the Historic House Hotels group and run as a hotel. Its proximity to Chequers means that it has frequently been the host of international and government summits and meetings.

The house has a remarkable history: its most famous resident was the Count of Provence, the exiled brother of the guillotined French King, Louis XVI. For five years from 1809 he and his wife, Marie Josephine of Savoy, lived there in a combination of great state and total chaos, with over a hundred courtiers and servants, who kept chickens and rabbits on the roof. Hartwell was acquired by Ernest Cook in 1938 and used as an army billet during the Second World War, subsequently becoming The House of Citizenship, a finishing school and secretarial college which remained in occupation until 1983. (*Historic England – Hartwell House, 1997*)

The house is surrounded by an 18th-century landscape

Hartwell House Bridge (© National Trust Images – Hugh Mothersole)

63

Hartwell House Bridge in 1960 (© Buckinghamshire Archives, 2020)

park and pleasure grounds naturalised by Sir Thomas Lee, third Baronet of Hartwell, in the style of 'Capability' Brown but completed by Brown's follower Richard Woods (1715–1793), one of the best-known landscape designers of the day. It is worth noting that Brown himself seems not to have worked at Hartwell, although some older books and even one current website claim that he did.

In 1900, the boot-shaped lake in front of the house was divided at its ankle by the central arch of a salvaged 18th-century stone bridge. Designed by James

The balustrades (© National Trust Images)

Paine, it had been the road crossing over the Thames at Kew from the 1780s until increasing traffic required a wider bridge. It was dismantled and sold at auction to the Lees in 1898. The Lees installed this central span of old Kew Bridge between the two lakes at the end of the 19th century, replacing Henry Keene's 18th-century bridge.
(Mawrey, 2018)

Detail of the vermiculated stone (© National Trust Images)

The carriageway (© National Trust Images)

One of the wing walls and its pillar (© National Trust Images)

The bridge itself is constructed from ashlar grey stone with a single graceful arch spanning 8 metres. The wing walls are splayed with a short square pillar at each end. There are balustraded parapets and the 3.5 metre carriageway is covered with gravel.

26 Thame Bridge

Bridge Name:	Thame Bridge (Bypassed and Disused)
Location:	Within Thame along the old B4011 towards Long Crendon. Right behind St Mary's Church on the old Thame Road. When the A418 bypass was built around Thame the old road out of the town in the direction of Long Crendon was closed to traffic and it is now a quiet broad foot path.
National Grid Reference:	SP 70354 06507
Crosses:	River Thame
Span:	9.5 metres
Bridge Type:	Iron Beam
Materials:	Iron and Masonry
Traffic:	Pedestrian
Opened:	1896 (current reconstruction)
Managed by:	Transport for Buckinghamshire
Historic England Designation:	Unlisted

Information: Although no longer used for traffic, Thame Bridge is accessible for pedestrians and dog walkers using the low-lying parkland which surrounds the area. There are three bridges identical in construction but varying in scale. The largest and most southerly (Thame Bridge itself) crosses the River Thame and the two smaller versions to the northeast traverse flood pools. The bridges are constructed on a decking of iron beams with wrought iron railings with black brick abutments and small stone towers. The bridges are currently in a stable but unsightly condition.

The boundary marker on Thame Bridge

The middle bridge – Thame

The iron railings

The Thame river forms the boundary between Buckinghamshire and Oxfordshire for a few miles between a point just north of the town and a mile or so to the west of Ickford. There is a county boundary marker on the bridge marking this border. There are parish records from 1309 documenting that Bishop Dalderby granted an indulgence for its repair. After its destruction by floods in 1894, the bridge was reconstructed in 1896 at a cost of £4,600 and this is the bridge we see today. *(British History Online – Thame, 2001)*

Thame Bridge – west side

The tower with the benchmark, 3 metres into Buckinghamshire (© Farrow, 2014)

Benchmark on one of the Thame Bridge towers

Thame Bridge railings and towers

The causeway above the flood plain at Thame Bridge

Bridge Name:	Ickford Bridge together with Whirlpool Arch Bridge
Other Names:	Sometimes referred to locally as the 'S-bridges' because of the reinforcements.
Location:	Ickford near Aylesbury 4 miles northwest of Thame
National Grid Reference:	SP 64870 06470
Crosses:	River Thame
Span:	9.5 metres
Bridge Type:	Three Arch
Materials:	Stone
Traffic:	Pedestrian & Road
Opened:	1685 Inscribed on Ickford Bridge / Whirlpool Arch Bridge is 18th century
Managed by:	Transport for Buckinghamshire
Historic England Designation:	Scheduled Monument and Grade I Listed

Ickford, the main bridge

Ickford Bridge early 1900s (© Buckinghamshire Archives, 2020)

Information: The Thame river valley here is shallow and the land is almost flat, forming a large flood plain across which this bridge traverses. A bridge across the Thame existed on this site as early as 1237. In that year parish records show that Walter de Burgh was ordered to provide the keeper of Ickford Bridge with a great oak tree from Brill (Brohull) to produce wood for repairs. In that century, the bridge was variously recorded as *Wodebrugge* or *Widebrugge*.

The present bridge, which carries the road from Ickford to Tiddington, is actually two bridges within 25 metres of each other; Ickford Bridge and Whirlpool Arch Bridge. The westernmost Ickford Bridge over the River Thame is an ashlar stone structure of three elliptical arches, two larger ones with a span of about 3 metres each and a central supporting pier. The third and smaller arch in the western approach causeway acts as a flood arch.

The pedestrian recesses within the walls of the main bridge

The more easterly second bridge known as Whirlpool Arch Bridge, crossing a smaller side stream with a small widening 'whirlpool', is a single stone arch forming a smaller humpback structure. Both have been reinforced with iron tie bars with backward 'S's acting as retaining plates. *(Historic England – Ickford, 2007)*

On the main bridge the triangular cutwaters on each side of the northern pier continue upwards to the parapets and form small pedestrian recesses. In the recess on the east side of the bridge are two country boundary stones, the southern one inscribed:

1685, Here ends the county of Oxon,

… and the northern one:

Here beginneth the county of Bucks, 1685.

Ickford Bridge in winter 1990 (© Buckinghamshire Archives, 2020)

The deck/roadway of the smaller eastern bridge

The eastern Whirlpool Arch Bridge – Ickford

Detail of the rising cutwater and the retaining 'S' reinforcements

Vehicle damage in the parapets

The roadway of the main bridge – Ickford

There is damage visible in the parapets of both bridges, evidence of the toll that modern wide vehicles have inflicted.

In the field to the left (north) there are old earthworks thought to have been constructed in the Civil War to defend the river crossing.

Downstream from the bridge is the venue for the annual tug of war contest between Tiddington and Ickford. The contest started in 1953 and the losers end up in the river. *(British History Online – Ickford, 2001)*

Both bridges are Scheduled Monuments.

Ickford Bridge 1945. Both bridges can clearly be seen (© Buckinghamshire Archives, 2020)

PADBURY BROOK

Padbury Brook rises from its source just north of Fringford in Oxfordshire. The brook heads eastward, north of Twyford, Buckinghamshire, continuing north of Steeple Claydon, where it is known locally as The Planks, to Oxlane bridge near Padbury. It meanders through the countryside and doesn't appear to have been significantly straightened. Padbury Brook then flows northeast in a wide and shallow valley passing northeast of Padbury village. The brook goes under the A413 road just north of Padbury, then under the A421 road (Buckingham to Milton Keynes road), then under Thornborough Bridge and on to Kingsbridge, where it is joined by its other arm, from the west (Thornborough) and becomes 'The Twins', before joining the River Great Ouse, east of Buckingham and west of Thornborough. The brook is joined by Claydon Brook, a major wandering tributary, which rises on the high ground on the southern edge of Buckingham.

28 Thornborough Bridge – FEATURE BRIDGE

Bridge Name:	Thornborough Bridge (Disused)
Location:	Thornborough Bridge is located on the original Bletchley to Buckingham road, now bypassed by a modern bridge built in 1974 for the A421. The bridge is accessible to walkers from an adjacent lay-by.
National Grid Reference:	SP 72925 33199
Crosses:	Padbury Brook
Span:	30 metres
Bridge Type:	Arch
Materials:	Coursed Rubble and Dressed Stone
Traffic:	Pedestrian
Opened:	Medieval (around 1400)
Managed by:	Transport for Buckinghamshire
Historic England Designation:	Scheduled Monument and Grade I Listed

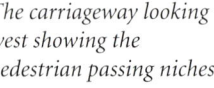

The carriageway looking west showing the pedestrian passing niches

Thornborough Bridge

Information: Thornborough was an important place in Roman times, marking the crossroads of five major roads. This east to west river crossing is at least 1,200 years old and dates back to Roman times. This position on Padbury Brook was originally a Roman ford, made from limestone blocks held in place with wooden stakes. These were found during excavations in the 1970s.

There are two large burial mounds to the north of the bridge that date to around AD 200, and when excavated in 1839 by the Duke of Buckingham, produced a wealth of treasures; bronze and gold ornaments, pottery, and glass. A glass urn containing human remains also survived, indicating that the mounds resulted from a lavish funeral ceremony.

The bridge itself dates to AD 1400 and is the sole surviving medieval bridge in Buckinghamshire and the oldest bridge in the county. It is made of stone with six stone arches, the centre two accommodating the main flow of the river. These centre arches are more ornate with ribbed vaults and carved voussoirs. Both arches are divided into three bays over the channels by two hollow-chamfered ribs and have hoodmoulds to the south side. The four remaining arches, two on each side, are less ornate and intended for spring flooding overflows. There are three cutwaters on the south side central piers which are carried up through the parapet as passing pedestrian places. The north side has one square passing place with an inscription tablet and shields carved onto the pier below.

The bridge straddles the parish boundaries of Thornborough and Buckingham. The parish boundary follows the line of Padbury Brook or The Twins, a tributary of the River Great Ouse. The parish division is marked by a boundary stone in the middle of the bridge. The stone bridge is around 30 metres long and 4 metres wide.

There is evidence of multiple repairs carried out over the centuries, including extensive repairs in 1661 and other repairs in the 19th and 20th century. There are fragments of an inscription on the north side of the bridge mentioning the 1661 repairs:

Thornborough Bridge – the 1661 repair inscription (© Buckinghamshire Archives, 2020)

This bridge was repaired by charge of the county ano domini 1661 according to a session order: sr r-hard f mp sr toby tyrel barron ts … and Thomas Stafford esquire being appointed to take order for the booking for the doing of the same and will … head of Thornborouye …of the said work Thomas … c…ssan

The parapets show 20th-century damage by modern vehicles struggling with such a narrow road width.

Detail of the eastern arch

The cutwaters

Thornborough Bridge in winter

The northern face of Thornborough Bridge

Repairs on the north side

More recent 19th-century repairs

Modern early 20th-century damage to the parapets

Detail of the less ornate western flood arch showing the vaulting

The carriageway – Thornborough Bridge

29 Oxlane Bridge

Bridge Name:	Oxlane Bridge
Other Names:	Oxlade Bridge
Location:	Just west of Padbury on Oxlane Road
National Grid Reference:	SP 71029 30305
Crosses:	Padbury Brook
Span:	32 metres
Bridge Type:	Four Arches
Materials:	Red Brick with stone cutwaters
Traffic:	Pedestrian & Road
Opened:	The original bridge at this point predates 1722
Managed by:	Transport for Buckinghamshire
Historic England Designation:	Unlisted

Information: Oxlane Bridge carries what is today a local B-road over Padbury Brook west of the modern Padbury Bridge and the A413. It is constructed of red brick with stone cutwaters and brick parapets. It has four arches and under normal flow conditions the brook only flows through the centre arches, the other two smaller arches to either side being flood arches.

Oxlane Bridge

The roadway at Oxlane Bridge showing the parapets

It is a popular fishing location and a small wooden footbridge on the west side of the bridge allows foot access to the riverbank.

Historically this was the main route to Aylesbury via Winslow and Whitchurch. The new A413 is comparatively recent, being a brand-new route for a turnpike, created around 1722. For hundreds of years, travellers to Aylesbury and London, including those on the Buckingham 'Old' Stagecoach, took an entirely different route over this bridge and some

Detail of the parapet capping and cutwaters

Oxlane Bridge 1979 (©Transport for Buckinghamshire)

parts of the journey included roads dating back to Roman times, or earlier.

Although still popular with local traffic, in 1937 the new A413 opened and the bridge and road became much quieter.

30 Claydon House Bridge

Bridge Name:	Claydon House Bridge
Location:	Middle Claydon
National Grid Reference:	SP 71748 24991
Crosses:	Private water fed by Padbury Brook
Span:	32 metres
Bridge Type:	Two Arches
Materials:	Stone and Masonry
Traffic:	Pedestrian (now abandoned and fenced off)
Opened:	1765
Managed by:	National Trust
Historic England Designation:	Grade II Listed

Information: The Claydon Estate lies in the Vale of Aylesbury, on low clay hills, at the centre of the four Claydon villages, 9 miles northwest of Aylesbury and 5 ½ miles south of Buckingham. Claydon House has been the ancestral home of the Verney family since 1620. The present Verney family, who still live on the estate, are the descendants of Sir Harry Calvert, 2nd Baronet, who inherited the house in 1827. The house was given to the National Trust in 1956 by Sir Ralph Verney, 5th Baronet.

View of Claydon Estate bridge looking northward towards the house

The park was landscaped by James Sanderson of Reading between 1763 and 1776. The brick-built bridge lies at the extreme southern end of the estate. It is unused and has been fenced off for many years, appearing today only as a grass covered hump in the landscape. Originally it formed a part of the southern

The overgrown bridge carriageway

drive into the estate, but it was bypassed (and forgotten) when the modern road to Botolph Claydon (Orchard Way) was created. Today the bridge sadly sits a mere 10 metres from the new road and goes unnoticed by the hundreds of daily road users.

The northern side of the bridge that faces Claydon House

It has a stone face to the north towards Claydon House with humpback brick parapets and rounded brick coping. The south side is constructed of English bond brickwork. It has a single round arch of only 1.5 metre span. On each end of the parapets there are square section ashlar piers. Probably erected or incorporated into the park landscape designed by James Sanderson 1763–1776 for Ralph, 2nd Earl, Verney. *(Historic England – Claydon, 1999)*

Detail of the southern side of the brick arch

Detail of the rounded brick capping. The modern road (Orchard Way) can be seen in the upper left corner of the photo

Bridge Name:	Three Bridges Mill (Bypassed and Disused)
Location:	Near Twyford
National Grid Reference:	SP 65627 27043
Crosses:	Padbury Brook
Span:	5.5 metres
Bridge Type:	Two Stone & Brick Arches
Materials:	Brick
Traffic:	Ramblers
Opened:	19th century
Managed by:	Transport for Buckinghamshire
Historic England Designation:	Grade II Listed

Information: There are indeed three bridges at Three Bridges Mill near Twyford, but the property has been abandoned except for a public footpath that runs through the site. The mill building remains at the far western end of the location. The bridge pictured here is the easternmost one and remains in the best condition of the three. A more modern bridge further to the east (not one of the three) carries a C road from Buckingham through Gawcott in the north towards Edcott, Twyford, and Steeple Claydon. There is a small layby on this road for walkers to access the site. A small industrial park for a wide variety of businesses occupies an adjacent area to the west.

The bridge is today a combination of a very old structure with extensive modern repairs and a disused roadway over Padbury Brook via two arches. The northern arch is of more modern engineering brick while the southern arch is of what appears to be original local stone. Both arches sit on a substantial central pier with a stepped brick cutwater. There are substantial abutments and unusual, stepped wing walls creating a causeway on both ends. There are no parapets, but more modern pipe guardrails have been ruthlessly added and one can see where incisions have been made into the substructure for modern concrete to support the pipes.

The stepped central cutwater

Three Bridges Mill eastern bridge

The stepped wing walls detail

The abandoned mill house

This bridge could well be the one referred to in a *Buckingham Advertiser* story of 22 January 1881. At the time of a great snowstorm a workhouse pauper perished and had an unfortunate journey to his grave. Despite the appalling weather, the Master of the Union Workhouse dispatched a porter with a hearse, and the corpse of the lad was to be taken for burial at Edgcott. The hearse became stuck on this bridge at Three Bridges Mill and it was only manoeuvred with great difficulty off the bridge to the Three Bridge Inn, where it remained in the care of the landlord for two days until the weather improved. Incidentally, the inn building which still sits on the other side of the bridge and is shown on OS maps as a public house now appears to be a private residence.

The older stone arch detail

Bridge Name: **Shipton Brook Bridge (Disused)**

Location: **Just south of Winslow on a bypassed section of A413 road. Signed as a picnic area.**

National Grid Reference: **SP 77654 26777**

Crosses: **Shipton Brook (a tributary of the Claydon Brook which is itself a tributary of Padbury Brook)**

Span: **31 metres including Abutments**

Bridge Type: **Three Arches – Single Lane with long causeways on each side**

Materials: **Black Brick**

Traffic: **Pedestrian & Road**

Opened: **1792**

Managed by: **Transport for Buckinghamshire**

Historic England Designation: **Unlisted**

Information: Shipton Bridge was built with a causeway across a flood plain in 1792, constructed out of black brick with low parapets. There is evidence of some major repairs over the intervening years. The bridge has three triple brick arches with brick cutwaters on the east side of the two centre piers.

Today the bridge sits on a little section of bypassed road (west side of the A413). This bridge now forms a lay-by with picnic facilities and the main road runs over a modern (1937) bridge a few metres to the

Shipton Brook Bridge west side

east. Shipton Brook bridge was built just south of Winslow for the new Aylesbury to Buckingham turnpike that opened in 1792. In 1937 the new bridge was built slightly upstream and the bridge was quietly bypassed.

This is a picturesque little bridge, but it is currently extremely overgrown making it difficult to appreciate.

The roadway of Shipton Brook Bridge

Shipton Brook Bridge

The milk tanker crash of 1932 (© Buckinghamshire Archives, 2020)

The photo to the left shows a crash involving a London Co-operative Society's milk tanker in 1932. The old bridge was inadequate for motor traffic. According to the *Buckingham Advertiser:* 'The tanker, which was going in a southward direction, arrived at the bridge at the same time as a large six-wheeled lorry, proceeding northwards. Crashing through the parapet of the bridge, the tanker fell on its side in the stream... Good fortune favoured the drivers, for neither was seriously hurt... There was a very considerable loss of milk, the stream being filled with many gallons of it.' (*Winslow History, 1999*)

In 1934, some skeletons were found by workmen in the sandpit at Shipton Bridge. Further investigation revealed the remains of at least four bodies. The jawbones have complete sets of strong teeth in them, and one appears to have belonged to a man of great strength. Another is that of a youth who had not yet cut his wisdom teeth. The police were informed, and a Home Office official was asked to investigate. The supposition is that these remains date back to Saxon days. *(Bucks Herald, 1934)*

It was also near this brook, now shrunk to very small dimensions, that a Winslow ghost was once said to regularly walk. There are still to be found people in the town who can remember being terrified as children by the tales of this spectre, who was said to have the extremely unpleasant habit of carrying his own entrails in a sieve. No explanation of his unpleasant manners appears to be forthcoming, but he seems to have been a harmless creature, and was familiarly known – with that simple directness, characteristic of rural England – as 'Old Sieve Guts'. He seems to have retired from business when the 'Jubilee Cottages' were built, nearly 50 years ago. Perhaps Victorian architecture was too much for him. *(Winslow History, 1999)*

SHIPTON BROOK.

Shipton Brook and bridge around 1914. (© Buckinghamshire Archives, 2020)

Along the riverbank

One of the arches and cutwater – Shipton Brook

Mid-Buckinghamshire Bridges

RIVER CHESS

The River Chess is a chalk stream that rises just north of Chesham in the Chiltern Hills. Most of it has been covered over and culverted under Chesham itself but it emerges again on the south side of the town and runs through the beautiful Chess Valley before entering Hertfordshire at Rickmansworth, where it becomes a tributary of the River Colne. Historically, the clear chalk stream water of the River Chess, together with the fertile land, was ideal for growing watercress, and this industry flourished in both Chesham and Rickmansworth in the Victorian era. However, due to excessive water extraction the river has been prone to drought in recent decades and the cress industry has all but disappeared. Further downstream the river flows below parkland landscaped by 'Capability' Brown at Latimer House and the site of a 1st-century Roman villa close to the village of Latimer.

33 Chesham Town Bridge

Bridge Name:	Chesham Town Bridge
Location:	Germain Street in Chesham
National Grid reference:	SP 95874 01353
Crosses:	River Chess
Span:	3.5 metres
Bridge Type:	Arch on north side and concrete lintel beam on the south side
Materials:	Red Brick and Concrete
Traffic:	Pedestrian & Road
Opened:	19th century
Managed by:	Transport for Buckinghamshire
Historic England Designation:	Unlisted

Information: The town of Chesham stands on the river Chess, near its head, 2½ miles northeast of Amersham. The river at Chesham particularly impressed Thomas Baskerville, who in 1671 stayed at the 'Crown' in Amersham, which dated back at least to the century before, and may have occupied the site on which the present 'Crown' stands. He says of Chesham:

Here also runs a nimble stream with mills on it to grind meal for London, and in a room over the market house people are much employed to hoult, cleanse, or sort the flour from the bran.

(British History Online – Chesham, 2004)

Contrary to popular belief, the town is not named after the river; rather, the river is named after the town.

There have been a number of bridges at this site since the 15th century and

Chesham Bridge pre-bypass in 1960 (© Buckinghamshire Archives, 2020)

the current structure is a more modern replacement for these. It is constructed of brick and forms a low 'humpback' when viewed along with the abutments and wing walls. The parapets are capped with modern concrete coping stones painted alternatively an unattractive black and white. There are pipe handrails that line the riverbank on the west side accommodating a pleasant river walk.

From here the River Chess passes on southward where the land lies only 90 metres above ordnance datum, and leaves Chesham for Latimer parish. *(British History Online – Chesham, 2004)*

Chesham Town Bridge

Chesham Town Bridge roadway

Modern plaque on the bridge

Chesham Bridge in 1909 looking along the River Chess (© Buckinghamshire Archives, 2020)

85

Bridge Name:	**Bois Moor Road Bridge**
Location:	**Chesham**
National Grid reference:	**SP 96394 00650**
Crosses:	**River Chess**
Span:	**2 metres**
Bridge Type:	**Single Arch**
Materials:	**Dark Red Brick**
Traffic:	**Pedestrian & Road**
Opened:	**1800**
Managed by:	**Transport for Buckinghamshire**
Historic England Designation:	**Unlisted**

Information: This small low-profile bridge opposite the tennis courts in Chesham has a main span of only 1.5 metres and has stood in this spot for over 200 years. The River Chess at this point runs southwards via a concrete aqueduct and thus there are hardly any wing walls at all. However, there is a small additional arch to the south side within a small wing wall added later. The roadbed is approximately 5.5 metres wide without designated pedestrian walkways. The visible parts of the structure are primarily parapets capped with rounded coping bricks and there is evidence of extensive modern repair on the north one. In the south facing side of the south parapet there is a date stone with nearly indistinguishable engraving, but the date of 1800 can just be made out.

The barely readable inscribed stone in the south parapet

Bois Moor Road Bridge showing the inscribed stone in the centre of the wall

Bois Moor Road Bridge – Chesham

Detail of the stone capping

Bridge Name: Stoney Lane Bridge

Other Names: Latimer Park Bridge

Location: Off Latimer Road turn north onto Stoney Lane which leads up to the Latimer House and Estate.

National Grid reference: TQ 00441 98565

Crosses: River Chess

Span: 5.5 metres

Bridge Type: Iron Beam

Materials: Black Brick and Iron

Traffic: Pedestrian & Road

Opened: 1898

Managed by: Transport for Buckinghamshire

Historic England Designation: Latimer Park is listed but the bridge is not specifically identified.

The parapets and stone capped end towers

Information: This bridge crosses the River Chess to the historic village of Latimer and the Latimer Estate. It forms part of the eastern edge of Latimer Park (Grade II listed – although the bridge is not mentioned in Historic England's listing details). The bridge has been rebuilt several times and the most recent version dates from 1898. It is constructed of black architectural brick sitting on an iron beam (currently painted an out of character light blue) of 5.5 metre span. The parapets of the same material have recessed decorative squares and stone coping and are held on each side between short towers, which have their own large stone caps. The deck/roadway is 7.5 metres wide with narrow pedestrian walkways. A stone inscription in the middle of the bridge states: 'BUCKS. CO. 1898'. There is a weir and dam created in the 1750s to the near north side of the bridge that forms

Stoney Lane (Latimer Park) Bridge

Stoney Lane Bridge carriageway

the 'Lower Water' of the estate. George Johnson in his *History of English Gardening* (1829) praised the design of Latimer Park as the work of Lancelot 'Capability' Brown (1716–1783) but there is no other evidence to support this.

The village of Latimer was originally joined with the nearby adjacent village of Chenies. Together they were anciently called Isenhampstead, at a time when there was a royal palace in the vicinity. However, in the reign of King Edward III, the lands were split between two manorial barons: Thomas Cheyne in the village that later became known as 'Chenies', and William Latimer in this village. Latimer village came into possession of the manor in 1326.

The adjacent weir to the west side

Detail of the stone inscription 1898

Bridge Name:	Chenies Hill Bridge
Location:	Off Latimer Road, Chenies
National Grid reference:	TQ 01521 98780
Crosses:	River Chess
Span:	10 metres total gap
Bridge Type:	Brick Arch
Materials:	Brick
Traffic:	Pedestrian & Road
Opened:	c.1898
Managed by:	Transport for Buckinghamshire
Historic England Designation:	Unlisted

Chenies Hill Bridge

Information: The River Chess splits ¼ mile north of here and this bridge covers the river water not diverted. The first bridge encountered when turning off Latimer Road onto Chenies Hill is the Dodds Mill bridge *(see Chenies Place Woodside entry)* where the water diverted upstream was used to power the mill. About 100 metres further along, at a curve in the road, is this second small bridge that sits comfortably in its landscape and dates to the same era as the mill bridge, around 1893.

This is a dark brick-built structure with two brick arches supported in the middle by a central pier with a cutwater on both sides. Each arch has a span of 2.5 metres. The brick substructure rises evenly to create matching parapets that gracefully curve outward. The deck/roadway is 5 metres wide.

The bridge was rebuilt in the 1970s and the rubble left in a pile by the riverbank. In the ensuing years it became overgrown but bits are still visible. A portion of the original parapet coping is shown here.

Chenies Hill Bridge c. 1968 (© Buckinghamshire Archives, 2020)

Southern side of Chenies Hill Bridge

Chenies Hill roadway showing the two parapets of the bridge

37 Chenies Place Woodside

Bridge Name:	Chenies Place Woodside (Two Bridges)
Other Names:	East Bridge and West Bridge
Location:	Off Latimer road turn north, first building on the right is Dodds Mill.
National Grid reference:	TQ 01536 98723 and TQ 01571 98730
Crosses:	River Chess
Span:	2 metres
Bridge Type:	Single Brick Arch
Materials:	Red Brick
Traffic:	Footbridge
Opened:	1893 (both)
Managed by:	Privately
Historic England Designation:	Grade II Listed

Northern side of Chenies Hill Bridge showing cutwater and arch

The West Bridge

Garden Designer – Chenies Place Woodside

Gertrude Jekyll is well known to many enthusiastic gardeners and landscape designers. She was held in high esteem by the gardening world earlier this century, when she was acknowledged for her extensive design work both independently and in partnership with Sir Edwin Lutyens, the architect. Their successful partnership, with each influencing the other, resulted in over a hundred Lutyens/Jekyll designs and greatly contributed to the English way of life.

Gertrude Jekyll
(1843–1932)

Gertrude Jekyll's influence on modern day landscaping can be attributed to the firm principles she laid down regarding garden design and planting schemes.

She believed passionately in the understanding of beauty within the natural landscape and strived to create it in her own designs. Her dedication, industriousness, and no-nonsense approach were all aspects of her personal philosophies about life which led to the creation of some memorable garden designs.

Gertrude loved gardening for dramatic effect and is best known for long herbaceous borders with colour schemes running from cold (white, blue) to hot (orange, red) and back to cold again. She had an artist's eye for colour and contrasted plant textures to great effect.

She was a formidable plants-woman, who experimented with plants in her own garden at Munstead Wood in Surrey before recommending them to anyone. She taught the value of domestic plants familiar to gardeners today: hostas, bergenias, lavender, and old-fashioned roses. Jekyll concentrated her design work on applying plants in a variety of settings, woodland gardens, water gardens, and herbaceous borders, always striving to achieve the most natural effect.

Any garden was treated as an entirety with sections within, but each part complementing the other.

Gertrude Jekyll became involved in gardening relatively late in life, having been instructed by doctors to abandon her main passions, painting and embroidery, due to severe and progressive myopia. She channelled her creativity into gardening, having a background of knowledge and love of the subject, which had developed since childhood.

Gertrude became a prolific designer, completing around 350 commissions in England and America, executed without leaving her home, but by often extensive correspondence with her clients.

Gertrude Jekyll's design principles were simple but effective. Architecture was the main framework of the garden, with hard landscape features to display certain plants. Her feeling for the correct use of materials in design, respect for craftsmanship and understanding and implementation of planting schemes has left a legacy for gardeners to enjoy today.

Information: These two nearly identical little footbridges lie within beautiful private grounds originally created as the Chenies Place Estate, but which have now been split between the upper Woodside House (off Latimer Road) and lower Dodds Mill (off Chenies Hill). The River Chess was split ¼ mile north of here to create a mill race providing a strong and steady flow to Dodds Mill, which sat on this site but is now converted to a private dwelling.

The grounds, gardens, and bridges were designed by Edwin Lutyens and Gertrude Jekyll. Gertrude Jekyll is well known to many enthusiastic gardeners and landscape designers alike. She was held in high esteem by the gardening world earlier this century, when she was acknowledged for her extensive design work, alone and in partnership with Sir Edwin Lutyens, the architect.

The pond court within the Chenies Place grounds leads north onto a brick footbridge designed by Lutyens in 1893, which is Grade II listed. The footbridge crosses the mill race at a right angle to the main path running along the north side of the stream. The bridge has a wooden coping, pairs of seats at either end, a small wooden balcony on the east side, and brick paving.

Northeast of this bridge lies a level lawn and some 100 metres downstream (east) is a second footbridge dating from the late 19th century to early 20th century. This bridge is also Grade II listed and in similar style to the west footbridge but without the balcony.
(Historic England – Chenies Place, 1999)

The West Bridge walkway

The grounds with the East Bridge in the distance

The West Bridge 'balcony'

East Bridge

93

Steps leading up to the East Bridge   *The East Bridge as it looked in 1909 (© Buckinghamshire Archives, 2020)*

Chenies Old Mill 1980 (© The Lutyens Trust, 1999)

East Bridge walkway

Detail of the arch and keystone. Both bridges are identical in this respect

Architect

Described by Historic England as 'One of the greatest architects the country has ever produced', Edwin Landseer Lutyens (1869–1944) produced some of the most iconic country houses of the 20th century. Working for clients who spanned the worlds of banking, commerce, and manufacturing, he reflected an era that took the best from the past and adapted it to the needs of modern life.

Lutyens is widely held to be our greatest architect since Sir Christopher Wren. Before the First World War his reputation rested on his designs for country houses, moving from Arts & Crafts-style country houses to a Neo-Georgian manor. He had set up his architectural practice when he was just 19 years old and was fortunate to meet the celebrated garden designer, Gertrude Jekyll, who introduced him to her wide circle of friends, many becoming his clients. Lutyens became the go-to architect for wealthy Victorians and Edwardians.

As his fame grew, he was appointed in 1912 to design New Delhi, the new capital of India. Soon after the war ended in 1917, Lutyens was invited to join a working party advising on the treatment of the war dead on the Western Front and it was his creation of the Cenotaph in Whitehall that brought him to the attention of the world. He was appointed principal architect by the Imperial (now Commonwealth) War Graves Commission and in the years directly following the war, the majority of his projects were memorials to the dead and missing, both in England, and abroad.

Lutyens designed over 50 war memorials in cities, towns, and villages across England, including the Cenotaph in Whitehall. He also designed other commemorative monuments abroad such as the stark and sombre Memorial to the Missing of the Somme at Thiepval, France and the movingly austere cemeteries on the former Western Front.

RIVER MISBOURNE

The River Misbourne rises from its source in a field at Mobwell Pond, off the old Aylesbury Road, near the Black Horse Pub on the northern outskirts of Great Missenden. It flows for 17 miles down the Misbourne valley through Amersham and Denham to join the River Colne just north of where the latter is crossed by the A40 Western Avenue.

The Misbourne is a small river and passes through Little Missenden, Old Amersham, Chalfont St Giles, and Chalfont St Peter, and under the Chiltern railway line and the M25 motorway on its way down the valley. The river is a 'perch' stream, flowing over a bed of impermeable material on top of a porous substrate. The Misbourne is the only river completely contained within Buckinghamshire.

38 Missenden Abbey Bridges

Bridge Name:	Missenden Abbey Bridges
Location:	Within the grounds of Missenden Abbey at Great Missenden.
National Grid reference:	SP 89820 01028
Crosses:	River Misbourne (dry at this point)
Span:	1.25 metres
Bridge Type:	Single Arch (both bridges)
Materials:	Rubble Flint and Stone
Traffic:	Pedestrian
Opened:	Around 1790s
Managed by:	Privately
Historic England Designation:	Grade II Listed (both bridges)

Information: Missenden Abbey is a late 18th- to early 19th-century park, woodland, lake, and pleasure grounds, laid out around a country house on the site of a medieval Augustinian Abbey. Missenden Abbey was the first Abbey to be founded in Buckinghamshire and the first or second Arrouaisian (a French order of monks) house in England. That order was dissolved in 1583. The estate has been through a variety of owners who carried out substantial work to both house and grounds until 1947 when the Carrington family sold it to Buckinghamshire County Council. Today Buckinghamshire New University run several of their educational programmes there and have an interesting display of the estate's archaeology in their reception area.

The upper bridge and summerhouse showing the dry riverbed

Detail of the upper bridge spillway

Missenden Abbey estate lies at the head of the Misbourne Valley in the heart of the Chiltern Hills. The estate was much larger but, having been debated since the 1930s, in 1960 the Missenden bypass was constructed through the middle of the park, and now only a small road and bridge link the house and the western side of the park with the eastern side and the parish church. The bypass cuts the whole park in two and is a great intrusion into the site, both visually and physically.

A long drive is shown on the 1883 Ordnance Survey entering at the south end of the park, running roughly parallel to the River Misbourne. This drive (now disappeared) ran up to the south front of the house, crossing the still existing rustic bridge over the Misbourne in the garden.

The River Misbourne, running south from where it rises in the village, has been culverted beneath the north-eastern walled garden. The river emerges beneath a restored flint and brick, battlemented, three-arched Gothic summerhouse built in the early 1790s and is itself Grade II listed. The riverbed runs through a rubble flint arched bridge built at the same time and also Grade II listed. It meanders down to another overgrown stone flint rustic bridge (also built at the same time and Grade II listed), which carried the drive from the park to the south front of the house. This second bridge was meant to create a backwater that cascaded into the lower stream bed; however, the river seldom flows, and the riverbed is dry. South of the garden the river widens into the lake bed known locally as Warren Water, also dried up.

A pump in the carriageway of the lower bridge. It was supposed to bring up water when water was once there

The outflow of the second lower bridge where the water should be cascading into the stream bed

Summerhouse gable detail

Detail of the lower bridge spillway

Summerhouse keystone detail

Bridge Name:	Highmore Cottages Bridge
Other Names:	Mill End Bridge
Location:	Highmore Cottages Lane off the A413 towards Little Missenden.
National Grid reference:	SU 93429 98451
Crosses:	River Misbourne
Span:	1.5 metres
Bridge Type:	Single Brick Arch
Materials:	Brick
Traffic:	Pedestrian & Road
Opened:	c.1700 but rebuilt in 1978
Managed by:	Transport for Buckinghamshire
Historic England Designation:	Unlisted

Information: A picturesque little bridge which might go unnoticed were it not for the pond on the east side of the road. The bridge has low parapets with coping bricks and the abutments with wing walls also form the edge of the pond. Interestingly the Misbourne passes directly under the adjoining cottage, usually the sign that the structure was originally a mill but there is no documentary evidence to support this.

View along Highmore Cottages Lane looking northwest with the bridge

Highmore Cottages Bridge east side

View of the Misbourne as it disappears under the cottages

Highmore Cottages bridge in the 1960s (© Transport for Buckinghamshire)

Bridge Name:	Mill Lane Bridge
Other Names:	Upper Mill or Sibley's Mill
Location:	Old Amersham
National Grid reference:	SU 95324 97538
Crosses:	River Misbourne
Span:	2.5 metres
Bridge Type:	Three Brick Arches
Materials:	Brick
Traffic:	Pedestrian & Road
Opened:	c.1700
Managed by:	Transport for Buckinghamshire
Historic England Designation:	Unlisted

Information: This bridge is associated with a historical mill, one of the three mills in Amersham mentioned in the Domesday book. This mill used to be known as Upper Mill or Sibley's Mill, after the name of a 19th-century malt miller. The present building dates from around 1700 when the 'pitt wheel' and a new stone were put in. It was bought from the Cheyne family by William Drake in 1792. This mill was a working flour mill until the 1930s and was previously a paper mill and earlier used by the Wellers for their brewery. *(British History Online – Amersham, 2001)*

Today's bridge is somewhat silted and photos of the bridge through the years show that the river once flowed more fully. A letter in the local museum tells of a miller at the west end of town asking the squire for a rent reduction as the water was too low to work the mill. *(Amersham Museum, 2018)*

There have been many paintings over the decades, and several are shown here.

The old town mill with the bridge around 1700, oil on canvas – artist unknown (© Amersham Museum)

Mill Lane Bridge carriageway with right angled wing walls

The carriageway looking south

Mill Lane Bridge – Old Amersham. The house in the background is the old mill

The Amersham town mill with the bridge in 1950 (© Buckinghamshire Archives, 2020)

The town mill with the bridge around 1908 (© Amersham Museum, 1908)

The old mill in 1998 (© Buckinghamshire Archives, 2020)

The Amersham town mill (Sibley's Mill) with bridge 1930 (© Buckinghamshire Archives, 2020)

Bridge Name:	Priory Bridge
Location:	At the end of Rectory Lane in Old Amersham behind the public Remembrance Park.
National Grid reference:	SU 96009 97385
Crosses:	River Misbourne
Span:	2.5 metres
Bridge Type:	Single Arch
Materials:	Brick
Traffic:	Pedestrian
Opened:	Probably 1870/90 when the church was remodelled
Managed by:	Privately
Historic England Designation:	Unlisted

Information: This little bridge forms a section of the public footpath and was once a part of the priory attached to St Mary's Church. It is a single triple brick arch with parapets that have rounded coping bricks. The abutment on the northeast side turns at a hard-right angle and the parapet walls follow this. To the southwest the Misbourne is contained within a brick channel next to the public footpath.

In dry years the Misbourne dries up completely.

Priory Bridge in 1960 (© Buckinghamshire Archives, 2020)

The town mill with the bridge around 1890, watercolour – artist unknown (© Amersham Museum)

Priory Bridge with the cemetery in the background

A completely dry Misbourne

Detail of the wing walls with capping brick

The view along the Misbourne looking north towards St Mary's Church

Priory Bridge – Amersham

42 Village Road Bridge

Bridge Name:	Village Road Bridge
Other Names:	Misbourne Cottage (village) Bridge, 'Murder Ahoy' Bridge
Location:	Village Road, Denham off the A412 past Denham Village Hall.
National Grid reference:	TQ 04028 87043
Crosses:	River Misbourne
Span:	3.5 metres
Bridge Type:	Single Arch
Materials:	Red and Brown Brick
Traffic:	Road Traffic
Opened:	18th century
Managed by:	Transport for Buckinghamshire
Historic England Designation:	Grade II Listed

Information: In the heart of Denham village this little bridge has played a role in village life for centuries. Built of red brick, the span of the Misbourne is covered by a single 3.5 metre arch. It is Grade II listed by Historic England. The parapets are slightly offset from the substructure and are capped with rounded brick. The narrow carriageway is only 3.5 metres wide without dedicated pedestrian walkways. A more modern pedestrian footbridge has been built immediately to the east of the road bridge.

The bridge sits next to Misbourne cottage and they are often pictured together. Misbourne Cottage is a 17th-century Grade II listed structure, which was featured (along with the bridge and several other buildings in the village) in the 1964 film *Murder Ahoy* with Margaret Rutherford.

The river Misbourne flows west to east through the centre of Denham Village and the entire village has been designated as a conservation area. The three bridges over it within Denham are an important element in forming the special character of the area. The well-known 'picture postcard' view of Denham Village is taken from the western entrance to the village with the listed red brick humped-back bridge (Village Road Bridge) in the foreground.

Western side of the bridge

Village Road Bridge and cottage – Denham

Denham Village in 1900. The bridge can be seen in the distant left (© Buckinghamshire Archives, 2020)

Rounded brick capping on the parapets

The bridge and cottage as they looked in 1960 (© Buckinghamshire Archives, 2020)

The bridge and cottage in 1910 (© Buckinghamshire Archives, 2020)

The bridge and cottage in 1900. Tree in garden and no garage (© Buckinghamshire Archives, 2020)

Bridge Name:	Old Bridge at Denham Place
Location:	Within the gardens of Denham Place off the Village Road quite near Village Road Bridge in Denham Village.
National Grid reference:	TQ 03969 87037
Crosses:	River Misbourne
Span:	9 metres
Bridge Type:	Arch
Materials:	Stone
Traffic:	Pedestrian & Road
Opened:	7th century
Managed by:	Privately
Historic England Designation:	Grade II Listed

Information: Denham Place and its grounds are Grade II listed by Historic England. An existing house was rebuilt between 1688 and 1701 for Sir Robert Hill, Member of Parliament for Wendover, and High Sheriff of Buckinghamshire, with a surrounding landscape park, possibly by Lancelot 'Capability' Brown.

A 1695 oil painting by John Drapentier of the house and grounds shows how the estate sits adjacent to Denham Village with a lengthy red brick wall separating the two. The wall is over half a mile long and 3 metres high. It was built between the 17th and 19th centuries. It is also Grade II listed with a coping of vertically laid bricks raked to a point.

In 1742 the estate was inherited by the Way family, with whom it remained until 1920. The formal gardens were removed by Benjamin Way in the 1770s, except for the walled garden to the south and one of the ponds within it. It is possible that Lancelot 'Capability' Brown (1716–1783) who disliked formal gardens, influenced the new layout of the house as well as the gardens. A painting done in 1675 by Peter Hartover shows the entrance face of the house before its rebuilding by Sir Roger Hill from 1688.

Within the estate is a bridge with two semi-circular arches. It is made of brick with a fine stone statue, said to be of Neptune and possibly from Italy, standing in front of the central pier.

The estate was sold in 1980 and converted to offices.

This detailed 1695 aerial view of Denham Place, the country seat of Sir Roger Hill (1642–1729) MP and Sheriff of Buckinghamshire, displays the elaborate formal gardens popular in the late 17th century. Denham Village can be seen in the centre back, outside the wall, and the wrought iron gates are to the right of the village in this painting. The formal gardens in front of the house were removed in 1770. Public Domain (© Yale Center for British Art, Paul Mellon Collection)

This 1675 painting, of unusually large scale, depicts the entrance front of the great house of Denham Place, Buckinghamshire before its rebuilding by Sir Roger Hill from 1688 and has been attributed to the artist Peter Hartover. Public Domain (Collection of Gail Salomon, 20 Avenue Foch, Paris)

The grounds and drive, Denham Place with the famous wall on the right

Denham Place

The wall from the village side

A much less grand Denham Place in the 19th century

44 Old Mill Road Bridge

Bridge Name:	Old Mill Road Bridge Place
Location:	Off the A40 heading north just beyond Denham Village.
National Grid reference:	TQ 04444 86698
Crosses:	River Misbourne
Span:	1.75 metres
Bridge Type:	Arch
Materials:	Stone
Traffic:	Pedestrian & Road
Opened:	18th century
Managed by:	Transport for Buckinghamshire
Historic England Designation:	Unlisted

Information: This bridge is located on the eastern edge of Denham Village and sits in conjunction with Mill House, which was historically known as Town Mill as early as 1086 in the Domesday book. The Misbourne appears from under Mill House via a brick arch, runs across a picturesque garden area in its bed contained within low brick walls, and then under Old Mill Road bridge. The road bridge itself is placed on a curve in the road and has a span of only 1.75 metres covered by a single brick arch. It is constructed of red brick continuing from water level up to parapets with brick coping. The carriageway is approximately 6.5 metres wide with quite narrow pedestrian curbs.

The river as it flows from under Mill House

The Old Mill Road bridge at Denham

The carriageway, Old Mill Road Bridge – Denham

The Misbourne as it flows from under the old mill (now a house)

SOUTH BUCKINGHAMSHIRE BRIDGES

RIVER WYE

Travelling from north to south, the Wye flows for about 10.5 miles from West Wycombe through High Wycombe, Loudwater, and Wooburn Green to its confluence with the River Thames at Bourne End. It has two main tributaries, the Hughenden Stream, and the Wycombe Marsh Brook which both, like the Wye, are fed from freshwater springs that rise up through the Chiltern chalk.

The river's course takes it through West Wycombe Park, where it was incorporated into the 18th-century landscaping of the house, its gardens and estate, forming several watercourses, a lake, and a cascade.

The Wye then flows as a single watercourse through the predominately industrial area of western High Wycombe before reaching a small public open space in central High Wycombe near Westbourne Passage, where West Wycombe Road joins Oxford Road. Lords Mill, a corn/paper mill dating from 1717, and Ash Mill, a paper mill dating from 1596, were located along this section of the river, although there is little evidence remaining to indicate their actual location. At one time there were 37 water mills on or near the River Wye along the 11 miles from its source and Pann Mill in central High Wycombe is one of the few remaining, and the only one in partial operation.

Like Milton Keynes, High Wycombe has treated its native river particularly badly with a majority of the river covered with modern culverts and urban development as it passes through High Wycombe on its way to London. When it does emerge, there are a few modern footpaths and footbridges but there are only a few interesting or historical features worth mentioning. The Wycombe Society has, in recent years, begun a public campaign to de-culvert the Wye. South of High Wycombe the river splits with The Dyke running slightly to the south of the River Wye. The Dyke becomes the Back Stream which re-joins the River Wye in Loudwater. There are only five crossings of any interest.

45 West Wycombe Park Bridges

There are no less than six historic bridges that sit entirely within West Wycombe Park:

1. West Bridge
2. Rustic Bridge
3. Two Arch Bridge
4. Cascade Bridge
5. North Bridge
6. Walton Bridge

The 333-acre site lies 2.8 miles west of the centre of High Wycombe, where four Chiltern valleys meet. Most of the park runs along the western, upper part of the Wye Valley. West Wycombe House (c. 1710–1715, Grade I listed) sits within the park and has been the home of the Dashwood family since 1698. The house and gardens are owned by the National Trust. Visitors to the property can see most of the watercourses, although an old sawmill of the same architectural style (now Sawmill House) and its adjoining streams are within a private part of the estate.

The park has many small temples and garden buildings along with these six small bridges, set in a landscape that was designed by Thomas Cook in the manner of Lancelot 'Capability' Brown. The gardens are widely seen as an example of the Augustan style of gardens. Humphry Repton was consulted on the park's design in the 1790s but few of his ideas were implemented. The lake, in the form of a swan was created by damming the River Wye. It is the dominant feature in the pleasure grounds and is prominent in views northeast from the house, framed by these six flint bridges. There are several springs within the lake that in turn fill it with its characteristic chalk filtered water, producing its gin-clear appearance.

Five of the West Wycombe bridges are constructed of flint stone and sit within 200 metres of one another. The flint stone faces of each have been skilfully laid with a decorative pattern.

Rustic Bridge with grassy carriageway

Bridge Name: **West Bridge**

Location: **Within West Wycombe Park**

National Grid reference: **SU 82951 94440**

Crosses: **River Wye**

Span: **5 metres**

Bridge Type: **Arch**

Materials: **Rustic flint stone facia with brick vault**

Traffic: **Pedestrian**

Opened: **18th century (late 1790s)**

Managed by: **National Trust**

Historic England Designation: **Grade II Listed**

Information: The westernmost bridge (West Bridge) has a single arch spanning just over 5 metres. The River Wye here is shallow and wide. The abutments sit mostly in the river and the wings are earthen. The flintwork rises to form low parapets with stone coping. There are decorative urns placed on all corners of the parapets. The deck is a mere 2 metres wide and made of grass covered stone rubble.

West Bridge – West Wycombe Park

West Bridge carriageway

Detail of one of the urns

West Bridge parapets and urns

45 Rustic Bridge – West Wycombe Park

Bridge Name:	Rustic Bridge
Location:	Within West Wycombe Park
National Grid reference:	SU 83009 94425
Crosses:	River Wye
Span:	5 metres
Bridge Type:	Arch
Materials:	Rustic flint stone facia with brick vault
Traffic:	Pedestrian
Opened:	18th century (late 1790s)
Managed by:	National Trust
Historic England Designation:	Grade II Listed

Information: The next bridge to the east (Rustic Bridge) sits directly below the north lawn of the house. It is a narrow single arch bridge of the same materials and style, that sits beside a small dam creating a back water lined with wild mint. It is of a more rustic style than the other bridges, with flint coping and a carriageway of only 1.5 metres. It is unadorned and occasionally in the height of summer its bed is dry. This is supposedly the bridge in the opening scene of the 1986 film *Labyrinth*, where Jennifer Connelly runs across the bridge.

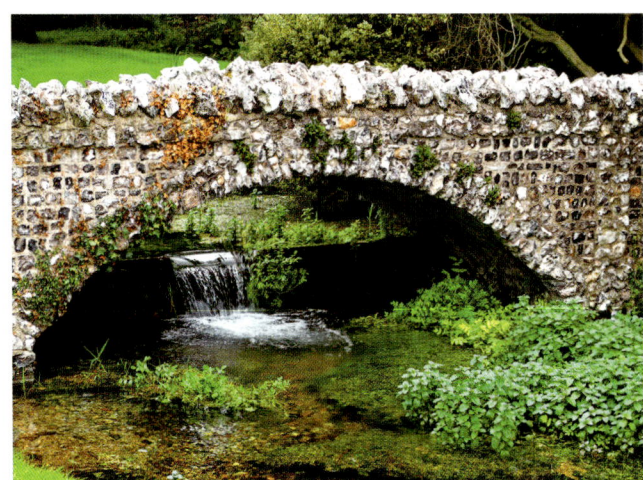

The arch and the weir

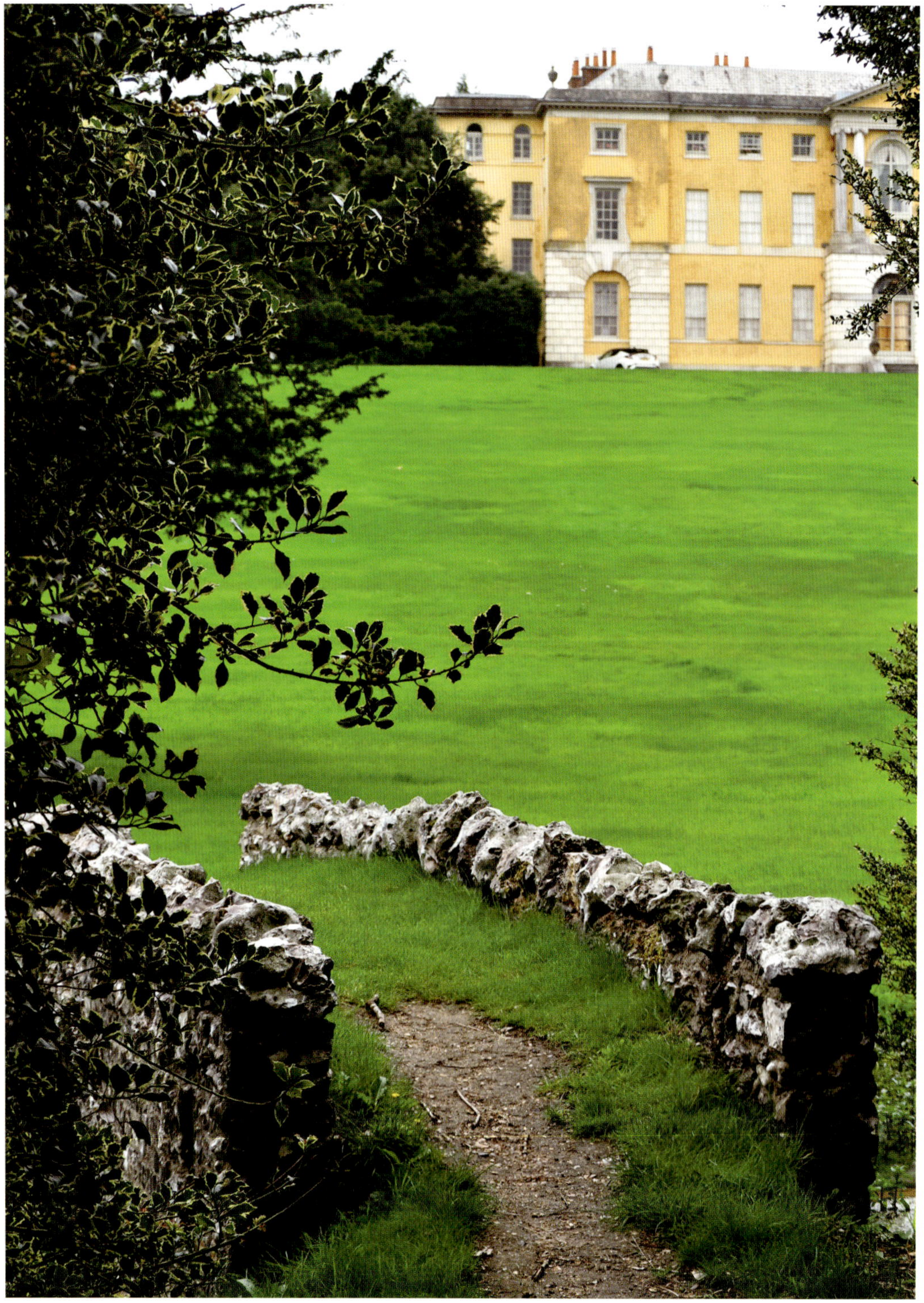

The bridge in relation to the house

Two Arch Bridge

45 Two Arch Bridge – West Wycombe Park

Bridge Name:	**Two Arch Bridge**
Location:	**Within West Wycombe Park**
National Grid reference:	**SU 83076 94380**
Crosses:	**River Wye (The Lake)**
Span:	**5 metres**
Bridge Type:	**Arch**
Materials:	**Rustic flint stone facia with brick vault**
Traffic:	**Pedestrian**
Opened:	**18th century (late 1790s)**
Managed by:	**National Trust**
Historic England Designation:	**Grade II Listed**

Information: Further yet to the east Two Arch Bridge is larger with a total span of 9.5 metres covered by two arches sitting on a central pier of the same material and architectural style. This bridge is the most formal of all the bridges in the park yet maintains its rustic flint stone charm. It has decorative offsets in the stonework and the two 3 metre arches have several substantial layers of flint. Again, there are urns on all four corners on the parapets and additionally two decorative spheres adorn the centres of the stone coping just above the piers. The 2.5-metre-wide deck is grass covered. This bridge connects a small island within the lake and the boathouse to the riverbank.

Two Arch Bridge showing the Hellfire Caves in the background

The urns decorating the coping

Bridge Name:	**Cascade Bridge**
Location:	**Within West Wycombe Park**
National Grid reference:	**SU 83231 94388**
Crosses:	**River Wye (The dam that creates the lake)**
Span:	**5 metres**
Bridge Type:	**Arch**
Materials:	**Rustic flint stone facia with brick vault**
Traffic:	**Pedestrian**
Opened:	**18th century (late 1790s)**
Managed by:	**National Trust**
Historic England Designation:	**Grade II Listed**

Information: Continuing the circular walk along the Swan Lake side, we come to Cascade Bridge, another similarly constructed low profile footbridge across the dam and cascade. It is made of five very low arches, allowing lake water to flow under it and across a cascade into the lower river. There are no parapets or adornments on the bridge itself. However, the view of the bridge is enhanced on both sides by flint plinths with reclining female figurines. There is a sluice gate next to the bridge on the south side.

Cascade Bridge from below the weir

One of the statues adorning the plinths – Cascade Bridge

Cascade Bridge close up

45 North Bridge – West Wycombe Park

Bridge Name:	North Bridge
Location:	Within West Wycombe Park
National Grid reference:	SU 83071 94558
Crosses:	River Wye
Span:	5 metres
Bridge Type:	Arch
Materials:	Rustic flint stone facia with brick vault
Traffic:	Pedestrian
Opened:	18th century (late 1790s)
Managed by:	National Trust
Historic England Designation:	Grade II Listed

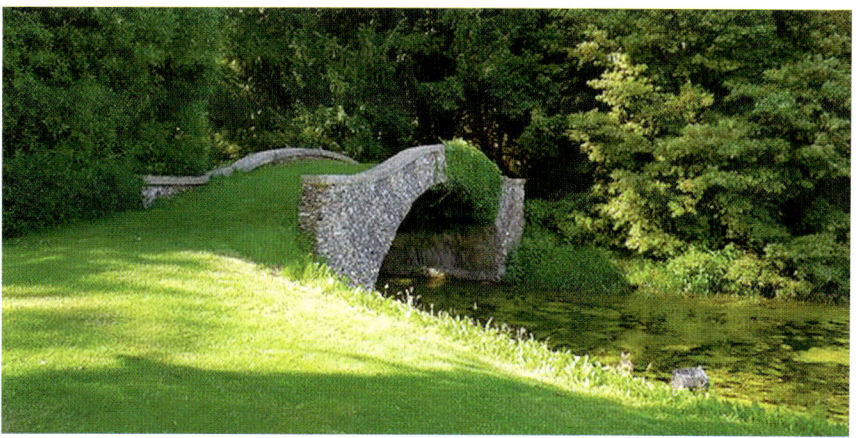

North Bridge in its setting

Information: Further around the lake is a true humpback bridge (North Bridge) that completes the circular walk around the swan shaped lake. It has a single span arch of 5.5 metres with thick stone coping blocks. This bridge is unmistakeably

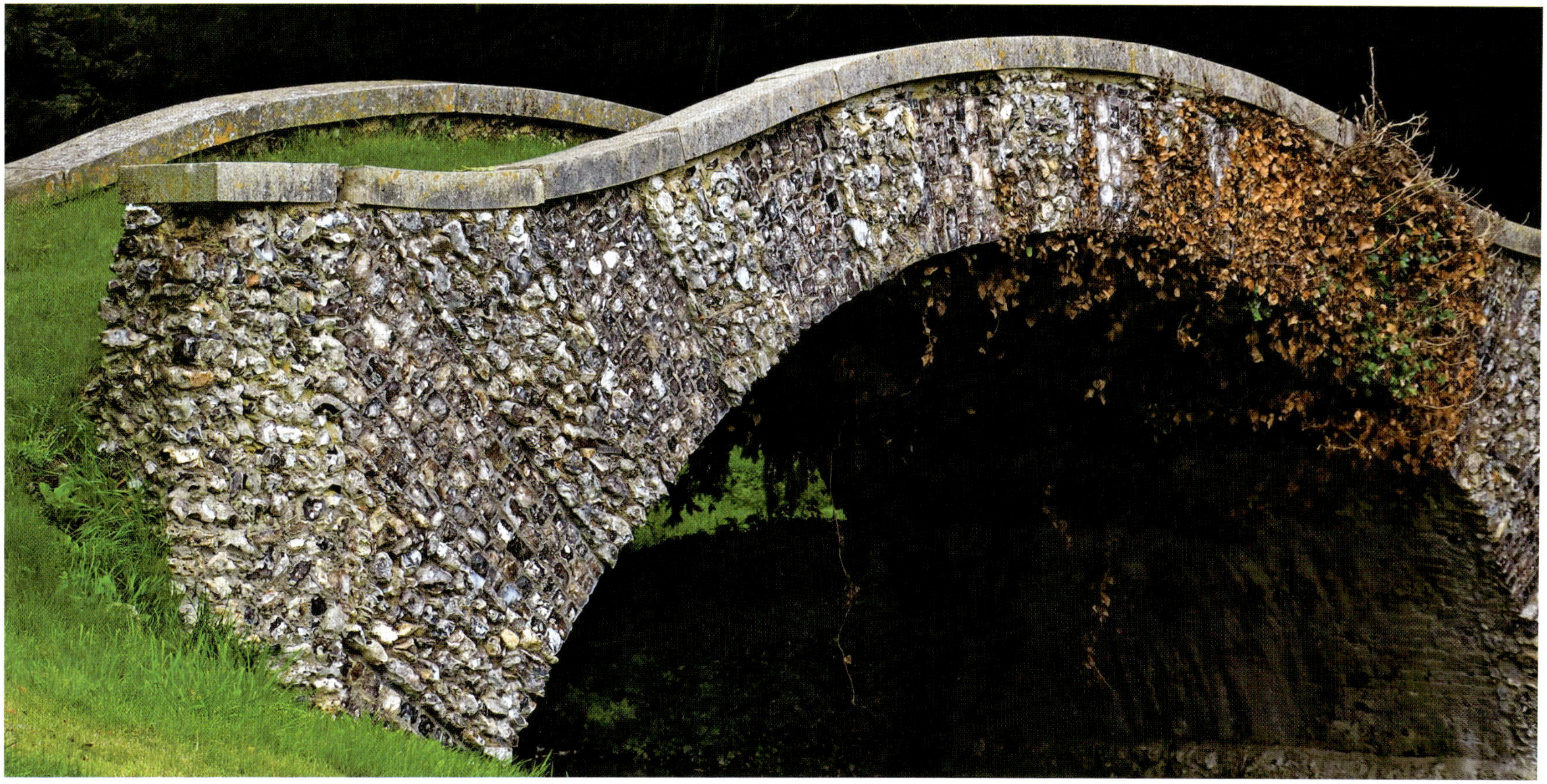

North Bridge

similar in style to Henry Repton's bridge at Stoke Park and perhaps this is his single design contribution. It has very low parapets and a short walk to the south will bring you back to the Rustic Bridge discussed previously.

Detail of the coping stones

The North Bridge carriageway

Bridge Name: **Walton Bridge**

Location: **Within West Wycombe Park**

National Grid reference: **SU 83544 94395**

Crosses: **River Wye**

Span: **3.5 metres**

Bridge Type: **Arch**

Materials: **Wood**

Traffic: **Pedestrian**

Opened: **18th century (late 1790s). Reconstructed in 2018**

Managed by: **National Trust**

Historic England Designation: **Grade II Listed**

Walton Bridge – West Wycombe Park

Information: The sixth and final bridge (Walton Bridge) sits alone, approximately 200 metres further east from the lake and is a modern replacement construction of wood for the original bridge depicted in the 18th-century painting and lithograph by William Hannon (1720–1772) also constructed of wood. It sits on flint and brick abutments with wooden handrails and this new bridge follows closely the design shown in Hannon's painting but without any adornment (urns) on the handrails.

These latter bridges are not mentioned specifically by Historic England but are themselves of 'listed' status by virtue of their location within the listed estate grounds.

The River Wye runs easterly on from the cascade in the west to the Pepper Box 'Pepperpots' Bridge (see p. 116) at the east end of the park. The Wye has been deliberately widened out in places to increase its visual impact.

In 1943 the house and park were given to the National Trust, in whose ownership they remain.

West Wycombe Park with Walton Bridge
William Hannan (1720–1772) (© Government Art Collection)

Walton Bridge – West Wycombe Park

Bridge Name: Pepperpots Bridge

Other Names: Historic England refers to this as 'Pepper Box' Bridge

Location: Chapel Lane, Sands, West Wycombe. The end of the West Wycombe estate.

National Grid reference: SU 84270 94000

Crosses: River Wye

Span: 6.5 metres

Bridge Type: Brick Arch

Materials: Brick and Flintstone

Traffic: Pedestrian & Road

Opened: 18th century

Managed by: Transport for Buckinghamshire

Historic England Designation: Grade II Listed

Pepperpots Bridge

Information: This very unusual bridge crosses the Wye at Chapel Lane in Sands. It has been listed separately from West Wycombe park by virtue of the fact that this is the only bridge affiliated with the park that carries a public road and has public access. This bridge appears at first impression as if it has, at some point in its life, been dissected lengthwise and reassembled in a way reminiscent of a Frankenstein creature. The west side of the bridge looks to be an intact part of the original structure. However, the eastern half looks to have been replaced with a more modern construction allowing the culvertisation of the Wye and the development of an industrial complex stretching south and east. The River Wye continues south-easterly but disappears underground at this point.

In actuality, this is the south-eastern end of West Wycombe Park and Pepperpots Bridge was created to complete the view for the Dashwoods, looking southeast from the park. This original bridge is interesting enough in its own right. Buckinghamshire County Council has this structure listed as a 'Garden Folly' affiliated with West Wycombe Park, and looking northwest from the bridge it does, in fact, provide a good view of the park and house.

Pepperpots Bridge North Tower

The carriageway deck is approximately 12.5 metres wide, accommodating two lanes of traffic and generous pedestrian walkways on both sides. The bridge is constructed of red brick and flintstone with parapets capped with rounded brick. The single low brick arch has a span of 4.5 metres and sits on stone imposts.

The abutments and parapets develop into lengthy walls of the same material and style, stretching both north and south for several hundred metres. Wycombe Farm, and Sawmill House, within the park are constructed of the same materials and in the same style, as are the dam and cascade that create the lake.

The distinguishing features of this bridge are the two tall towers at each abutment. Similarly constructed of the same flint stone and brick, they are about 3 metres square and 6.5 metres tall with roofs and chimneys. Doors lead to an unseen interior and the original rear windows have been filled in with modern concrete blocks. One can only guess at the original purpose of these towers but assumes that they are the origin of the 'Pepperpots' name. The entire edifice is overgrown and needs attention.

There are the ruins of a small chapel (perhaps a folly) of the same materials and construction methods in the woods to the north side of the bridge. There were low walls surrounding the chapel grounds that have now been reduced to rubble.

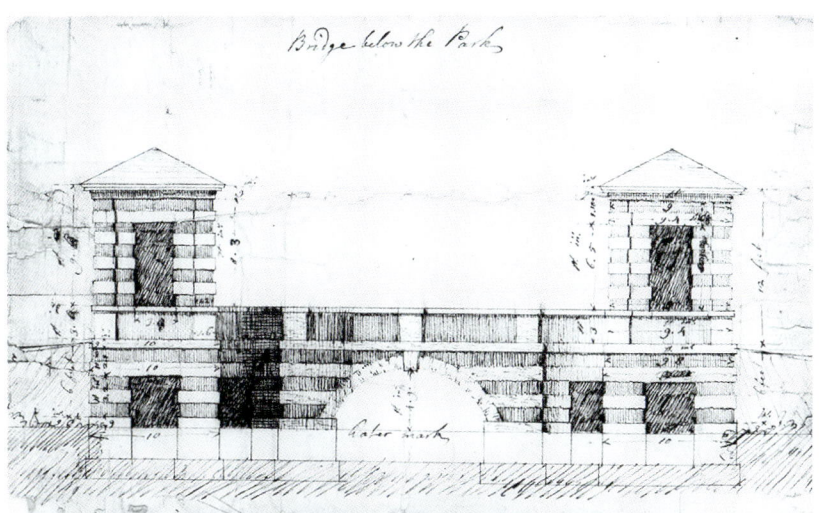

The original elevation in the architectural drawings – Pepperpots Bridge (Nicholas Revett, Design for a Bridge at West Wycombe 1778–1780, © National Trust)

Construction detail of the towers

Pepperpots Bridge looking southwest across the Wye

Pepperpots Bridge from river level showing the South Tower

Bridge Name:	Queen Victoria Bridge
Location:	Town Centre, High Wycombe
National Grid reference:	SU 86651 92802
Crosses:	River Wye
Span:	10.3 metres
Bridge Type:	Concrete Beam
Materials:	Concrete and Brick
Traffic:	Pedestrian & Road
Opened:	1901
Managed by:	Transport for Buckinghamshire
Historic England Designation:	Unlisted

Information: Unfortunately, the route of the Wye through the centre of High Wycombe cannot be seen until it emerges from a culvert at the back of the fire station by the Swan Theatre. From there it passes under Queen Victoria Bridge near the Police station to run for a short stretch in the small gardens by the side of the Council Offices, before submerging again under the A40 and reappearing close to the entrance to Pann Mill and The Rye public open space.

The ruins of the 'Chapel in the Wood' adjacent to Pepperpots Bridge

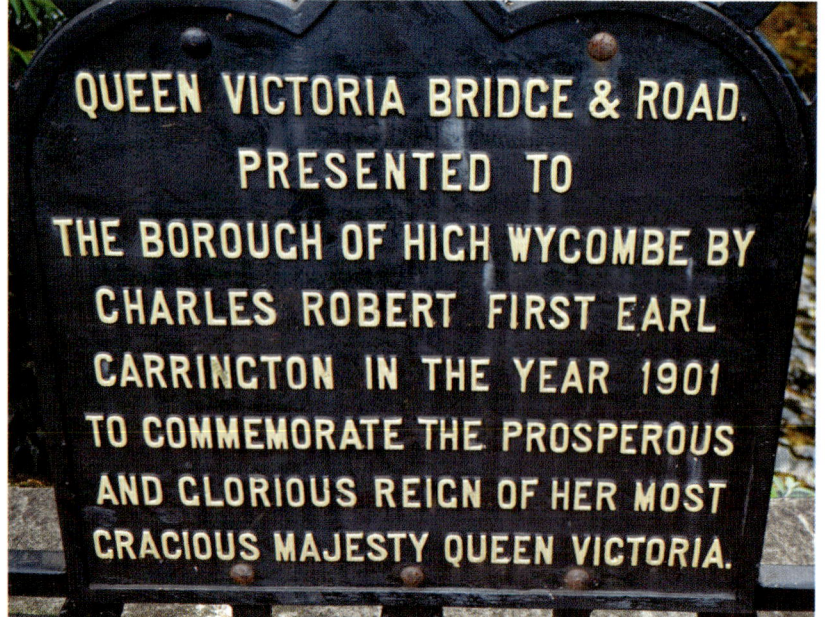
Queen Victoria Bridge High Wycombe plaque

Queen Victoria Bridge is not a particularly attractive or architecturally interesting bridge. It sits on a curved concrete beam resting on a brick and concrete pier with a rounded cutwater. The carriageway deck is a dual lane road which turns immediately into a roundabout. There are wrought iron railings supported by square pillars of red engineering brick capped with stone that add some visual interest. On the north railing there is a plaque which reads:

QUEEN VICTORIA BRIDGE & ROAD. PRESENTED TO THE BOROUGH OF HIGH WYCOMBE BY CHARLES ROBERT FIRST EARL CARRINGTON IN THE YEAR 1901 TO COMMEMORATE THE PROSPEROUS AND GLORIOUS REIGN OF HER MOST GRACIOUS MAJESTY QUEEN VICTORIA.

Queen Victoria Bridge looking southward. The Police station is on the immediate left of this photo

Queen Victoria Bridge looking northwards

48 Pann Mill Bridge

Bridge Name:	Pann Mill Bridge
Location:	On the south side of the A40 in High Wycombe.
National Grid reference:	SU 87304 92675
Crosses:	River Wye
Span:	1.5 metres
Bridge Type:	Concrete Beam
Materials:	Concrete and Brick
Traffic:	Pedestrian
Opened:	1759
Managed by:	High Wycombe Society
Historic England Designation:	Grade II Listed

Information: The first record of Pann Mill is in the Domesday census of 1086, and ownership of the mill changed many times over the years. An archaeological dig discovered that major rebuilding took place on at least three occasions. The oldest remains found dated from the 14th century. The most recent mill was built in 1759, with a new waterwheel and milling machinery fitted in around 1860.

The tiny bridge over the Wye forms part of what was once the narrow brick encased mill race running into the mill pond. There is still a sluice gate controlling the flow of water under a second bridge only 7 metres away. Both bridges are made of brick, forming low parapets sitting on concrete beams.

The mill now sits next to The Rye park space and is publicly accessible. There is a car park across the road.
(High Wycombe Society, 2019)

Pann Mill Bridge showing the mill race

The adjacent Pann Mill Bridge

Bridge Name: **Windsor Hill Bridge**

Location: **Windsor Hill Road, just east of Wooburn Green, between High Wycombe and Beaconsfield.**

National Grid reference: **SU 91511 88315**

Crosses: **River Wye**

Span: **5 metres**

Bridge Type: **Brick Arch**

Materials: **Brick and Stone**

Traffic: **Pedestrian**

Opened: **Early 19th century**

Managed by: **Transport for Buckinghamshire**

Historic England Designation: **Unlisted**

Information: This bridge is formed of a single span brick arch sitting on stone imposts. The brick face rises to form parapets on both sides. The deck/roadway is a narrow 4 metres, allowing only one vehicle to cross at a time and there are no pedestrian walkways. A modern steel guardrail protects the northern parapet.

Interestingly, the Wye enters this bridge from the north and them immediately after exiting on the south side, the river splits here for about a mile. It looks as if it may have been an intentional manipulation of the river's flow. The smaller diverted stream runs in a more westerly direction forced away from the large flow by an acute stone cutwater. This western arm then dives under a private driveway and proceeds to turn gently south again to eventually re-join the other branch.

There is a history of flooding in the area, the latest being in 2014.

The larger flow stream passes under a mysterious stand-alone brick wall with a similar brick arch allowing the stream to proceed to the south. This wall is approximately 3 metres in height with brick coping and is

Windsor Hill Bridge looking north

Windsor Hill Bridge roadway

Along the riverbank

supported on its western end by a wing wall, connected to a low wall, acting as a parapet for the private drive. On the wall's eastern end there is a rounded conical brick support pillar. The river moves here through a contained aqueduct suggesting that perhaps once upon a time this may have been the site of a mill *(undocumented)*.

Two scenes of the 1945 film *Blithe Spirit* were filmed on this bridge. In the first Margaret Rutherford (Madame Arcati) is summoned to the home of Rex Harrison (Charles Condomine) and on her way she is overtaken by Dr George Bradman (Hugh Wakefield) and his wife Violet Bradman (Joyce Carey) who have also been invited to the Condomine's. In the second scene filmed here Charles has had a fatal accident on this bridge and he joins his wives Elvira and Ruth as a spirit.

The point of division showing the acute cutwater

The conical wall support

The adjacent wall to Windsor Hill Bridge

HUGHENDEN STREAM

Hughenden Stream rises in the Hughenden Valley just east of Cryers Hill Road, and runs parallel with Valley Road for 2 miles, before disappearing into a culvert as soon as it reaches High Wycombe. It joins the River Wye, somewhere unseen, in a culvert near the town centre.

50 Hughenden Manor Bridge

Bridge Name:	Hughenden Manor Bridge
Location:	Hughenden Manor off Valley Road, the A4128.
National Grid reference:	SU 86616 95497
Crosses:	Hughenden Stream
Span:	2 metres
Bridge Type:	Stone Arch
Materials:	Stone
Traffic:	Pedestrian & Road
Opened:	1830
Managed by:	National Trust
Historic England Designation:	Grade II Listed

Information: This small stone bridge lies low in the landscape and forms an entrance route to Hughenden Manor and the surrounding grounds. Made of stone blocks with low retaining walls without parapets it is constructed in a gothic style, but dates to 1830, pre-existing the purchase of the estate by Benjamin Disraeli in 1847. The bridge was probably built during the ownership of the manor and grounds by the Savage family who held the estate from 1737 to 1847.

An archaeological study of the bridge has revealed a sequence of marks on the stones of the arch suggesting that the individual stones forming the arch were cut and shaped to fit elsewhere and were then brought to the site and assembled using the sequence of marks as a guide. The south wall carries a date in the stone of 1830 and it is likely that the marks originate from this date. A date of 1970 has also been observed in the cement mortar on the south parapet, suggesting when it was last repaired. No evidence for an earlier bridge or structure has been detected, suggesting this is the original stone bridge dating from 1830. The stream, however, would certainly have needed to be crossed and a drive is shown in this position as early as 1818. Prior to 1830 the stream may have been crossed by a ford, although a wooden bridge does seem

Hughenden Manor Bridge looking northward (© National Trust Images)

Hughenden Manor Bridge close up of the arch (© National Trust Images)

more likely. Perhaps the existing structure removed all trace of an earlier bridge, which could have been of timber construction.

In 1946 the estate was given to the National Trust in whose ownership it remains. The area that now forms Hughenden Park is owned and managed by Wycombe District Council. The entire estate and grounds are Grade II listed.

COLNE BROOK

The Colne Brook River splits from the River Colne at the A4007 at Uxbridge Moor and runs south along the Buckinghamshire border for several miles under the M25 and eventually crosses the Grand Union Canal. This marks the extent of its traversing Buckinghamshire. Proceeding south into Berkshire it again crosses under the M25 heading west, then under the M4 heading south.

51 Iver Bridge

Bridge Name:	Iver Bridge (Disused)
Other Names:	Chitty Chitty Bang Bang Bridge
Location:	Iver High Street, east side, just past Swan Road.
National Grid reference:	TQ 04120 81346
Crosses:	Colne Brook
Span:	12.5 metres
Bridge Type:	Three Segmented Arches
Materials:	Red Brick
Traffic:	Pedestrian & Road
Opened:	18th century
Managed by:	Transport for Buckinghamshire
Historic England Designation:	Grade II Listed

Information: This picturesque Grade II listed structure sits right next to Bridgefoot House (Grade II listed itself), not on the main road, but down a little private lane running north off Iver Lane. The bridge is constructed of red brick carried

Iver Bridge

The carriageway/deck of Iver Bridge

The wrought iron link insets within the capping stones

Two of the arches

Detail of the arch with retaining rod

continuously from the substructure through the parapets, which have thin stone coping with wrought iron link insets. Wing walls of approximately 4 metres lead to three brick arches without cutwaters. Extensive spalling of the brick is evident. Concrete reinforcement of the arches has been placed at water level. A little weir sits to the north of the bridge.

A couple of scenes from the 1968 movie *Chitty Chitty Bang Bang* were filmed here.

RIVER COLNE

The River Colne rises at Batchworth Lock just south of Rickmansworth where, for a short while, it runs as the Grand Union Canal before splitting off and running independently towards Moneyhill. It enters Buckinghamshire just east of Durden Court and continues south. For a short distance it runs parallel with the Grand Union Canal and passes under the A40. After this the River Colne passes out of Buckinghamshire heading south.

52 Iver Lane Bridge

Bridge Name:	Iver Lane Bridge
Other Names:	Clisby's Arch (Bridge)
Location:	On the B470, just east of Iver and west of Cowley, just before the Grand Union Canal bridge.
National Grid reference:	TQ 04579 81857
Crosses:	River Colne
Span:	8 metres
Bridge Type:	Three Segmented Arches
Materials:	Brick Arch
Traffic:	Pedestrian & Road
Opened:	Originally 1824 – rebuilt 1840
Managed by:	Transport for Buckinghamshire
Historic England Designation:	Unlisted

Information: This is a large bridge of about 50 metres in length. The substructure has a clear span of 8 metres crossing the River Colne that is traversed with a single six layered brick arch span of alternating spiral masonry work. The parapets of this bridge have a unique coping of cast iron forms filled with concrete. This type of coping is unique as far as I know. The central cap on the north side reads 'Thomas Harold Builder 1840'. The wing walls are finished at both ends with a circular brick tower also capped in the same style.

There are two coal tax markers at this bridge, one is a plaque on the north parapet and the other a post on the east end of

Iver Lane Bridge

The unique cast iron capping of the parapets The carriageway of Iver Lane Bridge

the north parapet. There are over 200 of these coal tax posts located around London, most of which are a metre high, white-painted bollards. The posts were erected under the Coal and Wine Duties (Continuation) Act of 1861 to mark the points at which duty had to be paid on coal being transported to London. The boundaries were originally set to be the same as those of the Metropolitan Police District, and the resulting revenues were used by the Corporation of the City of London to fund public works.

Coal tax marker post – Iver Lane Bridge

Detail of the 'skew' or spiral brick laying technique for bridge arches

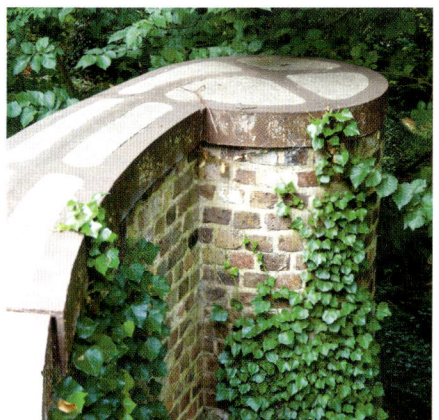

Detail of the cast iron capping on the circular brick tower

Detail of the central cast iron capping stone on the north side

Coal tax marker plaque – Iver Lane Bridge

Thorney Mill Road Bridge – the houses in the background were once the mill buildings

Bridge Name:	Thorney Mill Road Bridge
Location:	Just southeast of Thorney.
National Grid reference:	TQ 05399 79068
Crosses:	River Colne
Span:	14 metres
Bridge Type:	Four Arches, three of about 3 metres each and one smaller of 1.5 metres
Materials:	Brick
Traffic:	Pedestrian & Road
Opened:	1086 (reconstructed around 1962/63)
Managed by:	Transport for Buckinghamshire
Historic England Designation:	Unlisted

Information: Thorney Mill is where the River Colne is forced under what used to be the mill buildings via a series of four red brick arches. Probably one of the Domesday Mills. Because it was used for milling the arches are small and there is a substantial area of the span dedicated to creating a mill pond to increase pressure through the arches. The bridge itself forms the front of the mill and the parapet of the opposite side delineates the roadway. These buildings stood abandoned for many years until they were developed into housing during the 1980s. The wing walls of the bridge curve outward and force the river under the all-brick substructure. There are no cutwaters.

The mill ceased operations around 1903.

Thorney Mill Road Bridge carriageway

Detail of the arches and the retaining bars

FRAYS RIVER

Frays River splits from the River Colne just above Denham Lock. The River Misbourne flows into the Colne about 300 metres south of this point. The Frays runs east and south of the Colne for several miles. This appears to be the only section in Buckinghamshire and the remainder of the river is in Hertfordshire and Middlesex. The river disappears underground several times and crosses the Grand Union Canal again at Cowley Lock until eventually re-joining the River Colne east of Thorney. There is but one historic bridge that crosses the Frays River in Buckinghamshire.

54 Rockingham Road Bridge

Bridge Name:	Rockingham Road Bridge
Location:	The A4007 Rockingham Road in Uxbridge.
National Grid reference:	TQ 04566 83520
Crosses:	Frays River
Span:	10 metres
Bridge Type:	Three Arches
Materials:	Tan and Yellow Brick
Traffic:	Pedestrian & Road
Opened:	1809
Managed by:	Transport for Buckinghamshire
Historic England Designation:	Unlisted (Locally Listed)

Information: Rockingham Road Bridge is Locally Listed and dates from 1809. Local Listing indicates that the structure makes a positive contribution to local character and sense of place and offers some level of protection by the local planning authority, identifying them on a formally adopted list of local heritage assets.

This bridge has three brick arches with a dressed stone keystone above each arch. The substructure and superstructure are both made of the same yellow and tan brick without cutwaters. The level of the carriageway can

Rockingham Road Bridge – north side

Rockingham Road Bridge – south side

On the riverbank

clearly be seen from the side, and the brick parapets are capped with dressed stone. The width of the carriageway at traffic level is 11 metres and, oddly, there is a zebra pedestrian crossing at the height of the camber (in the middle) of the bridge.

There is a narrow strip of parkland to the north side and willow trees line the banks. A plaque in the middle of the north parapet reads:

THIS BRIDGE WAS FIRST BUILT IN 1809 AND WIDENED IN 1895.
THIS PLATE REPLACES A 5 TON WEIGHT LIMIT SIGN ERECTED AT THAT TIME.
THE ORIGINAL PLATE WAS REMOVED DURING THE 1939–45 WAR TO ALLOW
HEAVY MILITARY VEHICLES TO CROSS THE BRIDGE. 1996

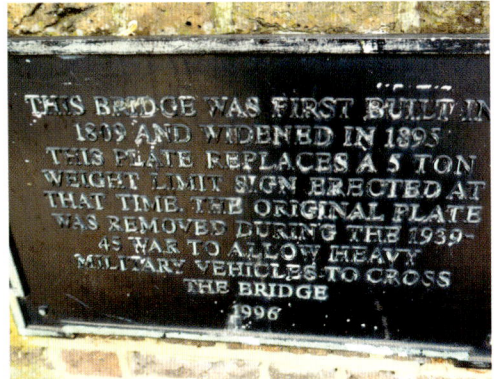

Detail of the plate on Rockingham Road Bridge

The crosswalk with the plate on the parapet in the background

Detail of the arch – Rockingham Road Bridge

STOKE PARK

55 Repton Bridge – FEATURE BRIDGE

Bridge Name:	Repton Bridge
Other Names:	Stoke Park Bridge
Location:	Stoke Park, Stoke Poges East Drive
National Grid reference:	SU 96700 82846
Crosses:	Private Water (now isolated)
Span:	11 metres
Bridge Type:	Three Arches
Materials:	Tan Portland Stone
Traffic:	Pedestrians & Golf Carts
Opened:	1789
Managed by:	Privately
Historic England Designation:	Grade II Listed

Information: Stoke Park was created in 1331 when Sir John de Molyns received a royal licence to enclose three woods. Over 400 years later the landscape park (now a golf course) in which this elegant bridge at Stoke Park sits, was originally laid out by Lancelot 'Capability' Brown around 1750, one of his first commissions. However, in 1790 when John Penn, who had inherited the estate from his father Thomas Penn, returned after 28 years in America, he found the manor house in a very bad state of repair and the grounds needing maintenance. He therefore set about building the present mansion between 1792 and 1808. He used much of the compensation he received from the new Commonwealth for the loss of his family's lands in Pennsylvania, following the American Revolution in 1776, to pay for it.

During that renovation in June of 1792 there were plans to alter the park and Humphrey Repton produced his ideas for improving the park landscape with a bridge with three arches. It was completed in 1798 and named after Repton who also redesigned the 'Capability' Brown landscape surrounding the house. Repton saw gardening as an art form with the landscape as his canvas, his ideal being 'natural beauty

The balustrades – Repton Bridge

Repton Bridge – Stoke Park

131

Repton Bridge carriageway

Repton Bridge west side

A local resident of Stoke Park

Detail of the north arch and balstrades

enhanced by art'. Repton outlined his principles of landscape gardening in his book *Observations on the Theory and Practice of Landscape Gardening.*

The white stone bridge, built by James Wyatt, is a graceful arc 14 metres in length with three stone arches, the central arch being slightly larger than the others. Each of the arches are approximately 2.5 metres in span. There are stone balustrades with square stone pillars at each end. It is Grade II listed.

The bridge was rebuilt between 2003 and 2005 as part of the Stoke Park restoration project. During restoration works several past repairs and alterations were identified, particularly from the 1940s when the structure was remodelled to accommodate heavy vehicles. The bridge was found to have a brick core, with Portland stone facing on the piers and ashlar limestone on the spandrels; much of the stone had been re-used from elsewhere. However, because Repton's thoughts were recorded in his 'Red Book' for John Penn and contained before-and-after sketches of parts of the estate, restoration work was carefully carried out over the last three decades to restore the fine bridge created by Repton.

These grounds are also famous for other reasons. It's where James Bond beat Goldfinger in possibly the most famous golf match in cinema history, where Bridget Jones ditched her singleton life for a weekend away with boss Daniel Cleaver, and where Teri Hatcher and Pierce Brosnan (as another Bond) danced the night away in *Tomorrow Never Dies.*

Landscape Artist – Stoke Park

Humphrey Repton was born in Bury St Edmunds in 1752. His passions were botany, entomology, and gardening and he worked to have a career in which he could enjoy these interests. Repton was determined to become a worthy successor to 'Capability' Brown. His landscape style was built on Brown's ideals, especially a vision of the house and how it was placed with relation to the

Humphrey Repton
(1752–1818)

landscape surrounding it. Repton saw gardening as an art form with the landscape as his canvas, his ideal being natural beauty enhanced by art. Repton developed further the landscape gardening style envisioned by Brown, introducing terracing as important to the foreground and gravel walks. He also reintroduced the flower bed and separate flower gardens.

Repton outlined his principles of landscape gardening in his book *Observations on the Theory and Practice of Landscape Gardening*. In this work he stated:

The perfection of landscape gardening consists in the four following requisites. First, it must display the natural beauties and hide the defects of every situation. Secondly, it should give the appearance of extent and freedom by carefully disguising or hiding the boundary. Thirdly, it must studiously conceal every interference of art. However expensive by which the natural scenery is improved; making the whole appear the production of nature only; and fourthly, all objects of mere convenience or comfort, if incapable of being made ornamental, or of becoming proper parts of the general scenery, must be removed or concealed.

Repton also published a book entitled *Sketches and Hints on Landscape Gardening* (1795). These books were based on his findings during his prolific career.

For every client Repton would keep a record in a red leather-bound book detailing each proposal for change, maps, plans, drawings, watercolours, and before-and-after sketches. These became known as his famous 'Red Books' – some of which survive, leaving a legacy of innovative ideas and helping to shape a new ideology which remains a part of modern-day landscaping practice.

Repton undertook over 400 commissions during his 30 year career, working on many important stately homes and gardens. Perhaps his garden most recognisable by Londoners today is Russell Square.

Image public domain

RIVER THAMES

I have seen the Mississippi. That is muddy water. I have seen the Saint Lawrence. That is crystal water. But the Thames is liquid history.

John Burns (1858–1943)

Compared to the giant rivers of the world such as the Nile, Mississippi, Amazon, or Ganges, the Thames is small, but what it lacks in size it more than makes up for with its unique wealth of history. From the very earliest times the Thames has played an important role in the development of Britain and Buckinghamshire.

The traditional source of the river at Thames Head, which is dry for much of the year, is marked by a stone in a field 108.5 metres above sea level and 3 miles north of the village of Kemble, southwest of the town of Cirencester in Gloucestershire. Some argue that a tributary, the River Churn, has a better claim to being the source; it rises near the village of Seven Springs 213 metres above sea level, just south of Cheltenham. There is also some dispute about the name of the river. Above Oxford it is still often called the Isis, and some old laws refer to the Thames as the Isis.

As it travels towards the sea the Thames works its way across a variety of rock and soil beds. It rises in chalk in Gloucestershire, then in Oxfordshire it cuts an easy path through clay before encountering flint, limestone, chalk again in Berkshire, and harder beds of gravel in Buckinghamshire. The river first created and then gave access to gravel and rock deposits, which made safe, firm foundations for crossings and eventually bridges. *(Lyte, 1980)*

Facts about the River Thames

- Length 215 miles
- The area of floodplain is 346 miles square
- There are over 100 bridges across it (more if you count footbridges)
- There are 47 locks
- The Thames has been frozen over at various times, the earliest recorded occasion being AD 1150.
- There is a 7-metre difference between low and high tide at London Bridge
- The Thames is navigable by barges for 191 miles from Lechlade
- The Thames is tidal as far as Teddington, approx. 68 miles
- The non-tidal part of the Thames from the source to Teddington stretches for 147 miles and falls some 104.2 metres
- 75 bridges cross over the non-tidal Thames
- 29 bridges cross over the tidal Thames
- It is estimated that the Thames carries some 300,000 tonnes of sediment a year

- More than 100 fish species have been recorded in the Thames estuary over the past 30 years, many of these in the river within London
- The country alongside the Thames (Thames Valley) is mostly rolling hills, with farming and grazing being the main uses of the land until London when it becomes urbanised.

Flow

The speed of the flow of the Thames increases the further downstream you go (towards the sea). This is because more and more tributaries join the river, adding their water to it.

- Buscot 799 million litres/day (176 million gallons per day)
- Reading 3,594 million litres/day (790 million gallons per day)
- Kingston 5,696 million litres/day (1,253 million gallons per day).

Width

The width of the Thames also increases the further downstream you go.

- Lechlade 18 metres wide
- Oxford 76 metres wide
- Teddington 100 metres wide
- London Bridge 265 metres wide
- Woolwich 448 metres wide
- Gravesend 732 metres wide
- Nore Light 6 miles wide
- Estuary (between Shoeburyness and Sheerness) 5 miles wide
- Whitstable and Foulness Point, the estuary is 18 miles across.

56 Cookham Bridge – FEATURE BRIDGE

Bridge Name:	Cookham Bridge
Location:	A4094, Ferry Lane in Cookham, Berkshire and Bourne End, Buckinghamshire
National Grid reference:	SU 89794 85612
Crosses:	River Thames
Span:	91 metres
Bridge Type:	Iron Beam
Materials:	Iron Girders & Brick Abutments
Traffic:	Pedestrian & Road
Opened:	1840 (Current structure 1867)
Managed by:	Transport for Buckinghamshire
Historic England Designation:	Grade II Listed

Information: Cookham Bridge carries the A4094 road across the River Thames above Cookham Lock. It connects Bourne End in Buckinghamshire with Cookham on the Berkshire bank. Prior to 1840 the river was crossed by ferry.

In 1837 the newly formed Cookham Bridge Company issued a prospectus expounding the 'great inconvenience and risk' of the ferry crossings and promoting the advantages of a quick and safe route over the Thames to access the Great Western Railway which was shortly to come to Maidenhead. The company proceeded with a design from George Treacher for a wooden bridge 88 metres long and 4.9 metres wide at an estimated cost of £2000. The bridge was wooden and had 13 spans, nine of 7.3 metres and four of 5.5 metres. This first Cookham Bridge was opened in 1840 in place of the ferry.

Due to its wooden construction the bridge required a lot of maintenance and only 19 years later in 1859 George Treacher, the original designer, reported to the Cookham Bridge Company that several of the piles were 'very much decayed and not unlikely to give way'. In Treacher's opinion

The wrought iron balustrades with wooden handrail

Manufacturing plate

View from Cookham Bridge – Stanley Spencer, 1936, Stanley Spencer Gallery (© Spencer, 1936)

Cookham Bridge looking north

The first Cookham Bridge as seen in 1865 before its demolition in 1867 (© Myers, 2016)

Cookham Bridge in 1874 (© Taunt, 1875)

Cookham Bridge around 1869 (© Myers, 2016)

Cookham Bridge deck/roadway looking north

the bridge was unlikely to survive the winter. *(Historical Cookham, First Cookham Toll Bridge, 2011)*

In 1866 the Cookham Bridge Company announced that a new iron bridge would be built and requested designs. Thirty-seven schemes were submitted, and the contract was awarded to Messrs Pease, Hutchinson & Co Ltd of the Skerne Ironworks, Darlington for a bridge of two wrought iron girders supported by eight pairs of concrete-filled iron pillars. The estimated cost was £2,520, some £1000 cheaper even than the estimate for the 1840 wooden construction. The remarkably low cost, due to Pease & Hutchinson being major iron manufacturers and prolific bridge builders, led to the new bridge being known as 'the cheapest bridge on the River for its size'. *(Historical Cookham, The Second Cookham Toll Bridge, 2011)*

Cookham Bridge, watercolour, 2001 (© Meyers, 2005) www.localartist.uk

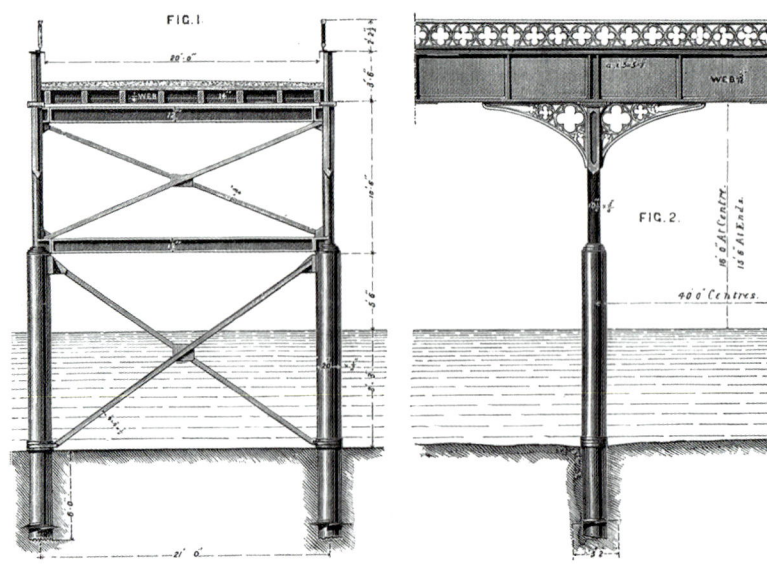

Cookham Bridge original architectural drawings (© Transport for Buckinghamshire)

Detail of the bridge supports – a good example of both lattice work and quatrefoil

Work on the replacement bridge began on 30 November 1866. The old bridge was demolished on 6 May 1867 and the approaches rerouted ready for the new bridge opening on 1 July 1867. The bridge continued to be owned by the Cookham Bridge Company and operated as a toll bridge until 1947 when it was bought out by Berkshire County Council for £30,000 and the tolls abolished. *(Phillips, 1981)*

The bridge has a single lane for vehicles controlled by traffic lights at each end, along with pedestrian pavements on each side. The vehicle weight limit is 7.5 tonnes, though buses and coaches are exempt from the limit.

The bridge is 102 metres long overall, while its length between abutments is 9 metres. The clear width of the roadway is 2 metres, and the height of the bridge at the centre from the bed of the river to the top of the handrail, is 9 metres. The bridge has a rise towards the centre of 76 centimetres. The superstructure consists of a wrought iron continuous girder supported at intervals of 12 metres. upon piers formed of iron piles. The main girders are 107 centimetres deep, with diagonal webs 13 millimetres thick, and top and bottom flanges 30 centimetres by 13 millimetres. The flanges of each girder are joined to the web by pairs of 76 millimetres by 76 millimetres by 13 millimetres angle irons. The girders are firmly bolted to the three centre piers, and rest on expansion

rollers. They also rest on expansion rollers at the two abutments ends. Ornamental quatrefoil cast iron brackets are introduced at the angles formed by the piers and girders.

The abutments are of red brick, with ashlar caps and a string course. Their face walls are 1 metre thick at the top, and the wing walls 66 centimetres thick at the top. The piers are each composed of two screw piles. To guard against the action of ice, the lower portion of each pile, up to about 1.5 metres above the water level, is made of wrought iron plate 13 millimetres thick. This lower portion of each pile is 32 centimetres in diameter outside, and it is filled in with concrete. The upper portion of each pile, which is of cast iron, is 26.7 centimetres in diameter and 19 millimetres thick. The transverse girders connecting the piles are 30.5 centimetres deep, and the cross bracing of the lower portions of the piles consists of 10 centimetres by 13 millimetres angle irons, and that of the upper portions of flat bars 10.2 centimetres by 19 millimetres. The piles are screwed down into the bed of the river to depths of from 1.8 metres to 3 metres.

The work was quickly and successfully carried out under the superintendence of Mr W. G. Fossick, Mr William Atkinson, C.E., of Westminster, acting as the representative of the Cookham Bridge Company. *(Myers, 2016)*

Over the decades since, Cookham Bridge has been the subject of many works of art and three are shown here.

Cookham Bridge © Timmy Mallett. Original oil painting by kind permission of the artist. No reprinting TimmyMallett.co.uk

Bridge Name:	Maidenhead Bridge
Location:	A4 Bath Road, Maidenhead / Taplow
National Grid reference:	SU 90141 81356
Crosses:	River Thames
Span:	13 Arches spanning 144.5 metres
Bridge Type:	Arch
Materials:	Portland Stone
Traffic:	Pedestrian & Road
Opened:	1280 (Current structure 1777)
Managed by:	Transport for Buckinghamshire
Historic England Designation:	Grade I Listed

Information: Maidenhead Bridge is a Grade I listed bridge carrying the A4 road over the River Thames between Maidenhead, Berkshire and Taplow, Buckinghamshire. It crosses the Thames on the reach above Bray Lock, about half a mile below Boulter's Lock. The Thames Path crosses the river here.

The town of Maidenhead grew up around the original wooden bridge, which was ordered by Henry III in the early part of the 13th century to ease the passage of travellers from London to the West Country. It forms part of the Bath Road, now the A4, which starts in west London.

The first bridge went up when Henry III issued a road widening order. That was to change the little village of Maidenhead forever. Suddenly it was on the road between London and Bristol, later to be called the Bath Road – now the A4. A timber wharf was built alongside the bridge and it is from this New Wharf or Maiden Hythe that Maidenhead takes its name; 'Maydenheth'. *(Maidenhead.net, 2000)*

From about the year 1250, the River Thames at Maidenhead was crossed by this piled timber bridge located slightly upstream of the present bridge site, but by 1297, the

Maidenhead Bridge

Detail of the northern flood arch through which the public footpath passes

South side of Maidenhead Bridge

structure was reported to be 'almost broke down'. Regular repairs were carried out between 1298 and 1428, though by 1452 it was said that traffic 'cannot cross without peril at certain times of the year through floods and the weakness of the bridge'.

The current Georgian Portland stone masonry road bridge was designed by Sir Robert Taylor (1714–1788), architect of the King's Works. In 1772, a Parliamentary Act was passed for its construction and a contract awarded to John Townsend of Oxford for £14,500. The foundation stone was laid in October the same year.

Opened to traffic in 1777 and built to the design of, and under supervision of Taylor, there are five arches over the river with four diminishing arches in each of the approaches. Each arch has projecting stone vermiculated voussoirs with a moulded cornice. There are moulded cornices and balustraded parapets across the entire length.

The bridge is 144.5 metres long and 9.1 metres wide (9.9 metres wide including its stone parapets) and is composed of 13 semi-circular arches. The arches are of Portland stone with rubble fill and vary in span from 10.7 metres at the centre to 8.7 metres at the sides. The parapet walls of the approach spans are plain.

The central arch was completed in 1775, and the bridge opened to traffic in 1777. The eventual construction cost recorded varies from £15,741 to £25,000.

In November of 1894, the Thames experienced severe flooding, swelling the river over 2.74 metres in height, nearly obscuring the cutwaters we can see today. An iron plaque can be seen on the northernmost cutwater indicating the flood mark.

Tolls were charged for crossing the bridge until 1903. In 1834, according to parliamentary records, the revenue amounted to £1,245 a year.

In the 1920s and 1930s this part of the river was very popular with members of high society who would visit Skindles Hotel, which stood on one side of the river and the Rivera Hotel on the other. Cliveden House, which was owned by the wealthy Astor family, is a short boat ride upriver.

In February 1950, the bridge was Grade I listed. Sir Nikolaus Pevsner (1902–1983) described it as 'Georgian masonry at its best'. It was cleaned and restored in 1977, and the brick soffits of the approach arches rendered in 1991–1992. *(Pevsner, 1991)*

Close up of the central arch and the balustrade

The vermiculated stone blocks making up the arches and parts of the piers

The flood mark plaque

In 2008–2009, repairs and strengthening works to the river spans included an upgrade of the foundations of the river piers to minimise underwater erosion. Eroded stonework was also removed and replaced. Traffic flow was unimpeded as the work was carried out from barges on the river.

Maidenhead Bridge continues to carry two lanes of traffic. A 40-tonne vehicle weight limit has been imposed. *(Engineering Timelines – Maidenhead Bridge, 2020)*

On the north side the public footpath along the Thames passes through the end flood arch providing a close-up view of the bridge's substructure. Like many of the bridges along the Thames, Maidenhead Bridge has been the subject of many artists.

Along the riverbank

Maidenhead Bridge, 2004, watercolour (© Myers, 2004) www.localartist.uk

Maidenhead Bridge

Detail of the cutwaters

Architect and Sculptor – Maidenhead Bridge

Sir Robert Taylor
(1714–1788)

Sir Robert Taylor (1714–1788) was born in Essex. His father, a London stonemason, apprenticed his son to the sculptor Sir Henry Cheere who sent him to study in Rome. He returned to England on receiving news of his father's death. Without an inheritance Taylor found himself penniless but began a career as a sculptor. The monuments to Cornwall at Westminster Abbey (1743–1746) and the figure of Britannia in the centre of the principal façade of the old Bank of England are his work as is the sculpture in the pediment of the Mansion House. The work at Mansion House was completed in 1753 and about that time Taylor gave up sculpture for architecture.

Through his connections as a sculptor, he came to be appointed as architect to the Bank of England, a position he held until his death, when he was succeeded by Sir John Soane. During his time there he was commissioned in 1776–1781, and again in 1783, to make substantial additions to the bank, which included the wings on either side of George Sampson's original 1733 façade.

Taylor also worked on Ely Cathedral, as well as several prominent country seats. Taylor was one of the three principal architects attached to the London Board of Works. He was surveyor to the Admiralty, and laid out the property of the Foundling Hospital, a prominent charity dedicated to the welfare of London's abandoned children of which he was a governor. He succeeded James 'Athenian' Stuart as surveyor to Greenwich Hospital, and was surveyor and agent to the Pulteney Estate (Bath) and many other large estates. He was Sheriff of London in 1782–1783, for which he was knighted. He died at his residence, 34 Spring Gardens, London, on 27 September 1788, and was buried in a vault near the northeast corner of the church of St Martin's-in-the-Fields.

In addition to Maidenhead Bridge, Taylor also designed six minor bridges on the Botley Road in Oxford (1767) (none survive), the Swinford Bridge over the River Thames at Eynsham (1767–1769), the removal of houses on the London Bridge and replacement of the central two arches by a single arch (1756–1766), demolished 1831, and he is credited with alterations to 10 Downing Street, London (c. 1780).

Image public domain

58 Marlow Bridge – FEATURE BRIDGE

Bridge Name:	**Marlow Bridge**
Location:	**Marlow**
National Grid reference:	**SU 85117 86102**
Crosses:	**River Thames**
Span:	**69.3 metres between pier centres, 112.5 metres overall**
Bridge Type:	**Suspension**
Materials:	**Steel & Stone**
Traffic:	**Pedestrian & Road**
Opened:	**23 September 1832 (Current Suspension Bridge)**
Managed by:	**Transport for Buckinghamshire**
Historic England Designation:	**Grade I Listed**

Marlow Bridge

Information: Marlow Bridge is the only suspension bridge in the county and carries road traffic and pedestrians over the River Thames. It is the third bridge to occupy this spot and the only suspension bridge over the non-tidal section of the River Thames. It connects the towns of Marlow in Buckinghamshire and Bisham village in Berkshire. It crosses the Thames just upstream of Marlow Lock, on the reach to Temple Lock.

There has been a bridge on the site since the reign of King Edward III. Documentary evidence of around 1530 records the bridge to have been of timber. In 1642 this bridge was partly destroyed by Major General Brown of the parliamentary army during the English Civil War (1642–1651). In 1789 a new timber bridge was built by public subscription with a contribution from the Thames Navigation Commission to increase the headroom underneath. However, by the early 1800s this bridge had fallen into disrepair. By 1828, it was reported that the second bridge had: 'fallen into a state of decay, so as to make the passage of it in a degree dangerous to the public'. (*Engineering Timelines – Marlow Bridge, 1999*)

The meagre revenues from local lands were insufficient for its maintenance and a court case decided the crown should 'provide and keep up a safe and proper bridge over the Thames at Marlow'. (*The Marlow Society, 2009*) Work began immediately on the current replacement bridge. The current bridge was built between 1829 and 1832, replacing the wooden bridge slightly further downstream.

Civil Engineer – Hammersmith Bridge and Marlow Bridge

William Tierney Clark (1783–1852)

William Tierney Clark was one of the most distinguished and widely known civil engineers in the first half of the 19th century. He spent all his professional life in Hammersmith. He was born in Somerset, Sion House, and as a youth he worked at Colebrookdale Ironworks, then the best place to learn the mechanics of cast and wrought iron fabrication for structural use. In 1811 the position of engineer became vacant at the West Middlesex Waterworks in Hammersmith, and Clark was duly chosen after an interview session.

By the early 1820s the need for a bridge at Hammersmith to shorten the journey to the Surrey side was becoming apparent. The Hammersmith Bridge Company was formed in 1823 and it accepted Clark's proposal for the building of a suspension bridge. Clark was strongly influenced, and the promoters reassured, by Thomas Telford's successful large span bridge over the Menai Straight of 176 metres. The foundation stone was laid inside the Middlesex cofferdam by the Duke of Sussex on 7 May 1825. The Hammersmith crossing was completed in four years and was the first suspension bridge over the Thames. It set the pattern for Clark's three subsequent bridges at Marlow, Shoreham, and Buda-Pesth (later Budapest). Count Széchenyi, the Hungarian aristocrat behind the Buda-Pesth Chain Bridge, was immediately captivated by the sight of Hammersmith Bridge and this led eventually to Clark's greatest challenge – building, or as he wrote 'throwing' the bridge over the Danube. A plaque commemorating Clark was installed in Budapest in 1998.

The building of Marlow bridge aroused great popular interest, particularly when the chains were raised. In 1829 the Boat Race was transferred to this part of the Thames and the bridge became a convenient viewing platform.

Clark also designed bridges in Bath and at Welbeck Abbey.

Image public domain

The current suspension bridge was designed by William Tierney Clark, with chains and structure designed by Samuel Brown. Clark had designed the first suspension bridge over the Thames, Hammersmith Bridge (opened 1827, replaced by the present bridge 1887). The smaller bridge at Marlow shares similarities with Hammersmith, notably in the use of masonry arches for the suspension towers. (*The Marlow Society, 2009*)

The main span of Clark's Marlow Bridge measures 69.3 metres between pier centres, with flanking spans of 21.6 metres to the north and 21.4 metres to the south. The pier bases are of rusticated masonry and brickwork, with Doric arches above deck level. The original carriageway is 6.1 metres wide with footways on either side of 1.5 metres wide.

In July 1949, Marlow Bridge was Grade I listed.

Over time the suspension chain ironwork in its anchorages corroded and several suspender rods fractured, thus the weight limit was reduced from 5.1 tonnes to 2 tonnes. The poor condition of the anchorages led to further plans for the bridge to be replaced, though local opinion was heavily in favour of retaining the existing structure. On 17 August 1962, it was decided to reconstruct the bridge to its original appearance, with the capability of bearing a 15.2 tonne vehicle. The weight requirement was later reduced to 5.1 tonnes.

The substantial restoration project was carried out in 1965–1966, by consultants Rendel, Palmer & Tritton, main contractors Horseley Bridge, and Thomas Piggott Ltd. Steel chains and a steel deck were installed simultaneously. The total cost was £223,000.

It was found that much of the wrought ironwork was still sound owing to the quality of the original workmanship. It was retained wherever possible and the necessary new steelwork replicated the original design. The palm tree motif on the heads of the handrail upstands is a new feature.

In 1972, a new balanced cantilever concrete bridge opened some 800 metres downstream (east) of the suspension bridge to carry the A404 trunk road.

On 24 September 2016, in the late evening, a 40-tonne lorry using GPS navigation attempted to cross the bridge and got stuck between the traffic calming measures on the north side. The bridge was assessed rigorously by structural engineers and found to be safe. It remains in daily use. (*Engineering Timelines – Marlow Bridge, 1999*)

Marlow Bridge south end showing The Compleat Angler

Marlow Bridge looking south from the pedestrian walkway

Marlow Bridge north tower

Keystone corbel on the north tower

Marlow Bridge from the air 1970 (© Buckinghamshire Archives, 2020)

Detail of the iron links

A postcard of Marlow Bridge around 1914 (© Buckinghamshire Archives, 2020)

The commemorative plaque at Marlow Bridge

The Chain Bridge in Budapest designed by William Tierney Clark

Bridge Glossary

A

Abutment – The structures in the ground at either end of the span of the bridge which carry the vertical and horizontal loads from the bridge superstructure. The ground ends support the bridge, especially to resist the horizontal thrust of an arch.

Arch – A curved or pointed architectural feature, symmetrical in form and supported at the ends. Used for spanning open spaces and capable of supporting the weight of a bridge superstructure above it.

Archaeological Interest – There will be archaeological interest in a heritage bridge if it holds, or potentially may hold, evidence of past human activity worthy of expert investigation at some point. Heritage assets with archaeological interest are the primary source of evidence about the substance and evolution of places, and of the people and cultures that made them. *(Historic England, 2018)*

Arch Bridge – A bridge with abutments at each end with a curved arch between them supporting the superstructure. Arch bridges work by transferring the weight of the bridge and its load partially into a horizontal thrust restrained by the abutments at either side.

Ashlar – Large squared blocks of any type of stone. Also frequently used for cut-stone masonry.

Assizes – Courts held in the main county towns and presided over by visiting judges from the higher courts.

B

Baluster – One of a set of vertical support shafts, posts that are topped by a rail.

Balustrade – A set of balusters and top rail when considered as a whole. Usually used to prevent people from falling over the edge of a bridge, balcony, or stairs.

Beam – Structural underpinning member that carries horizontal weight by transferring load to support structures at each end.

Beam Bridge – Bridges where beams, usually underneath but sometimes on

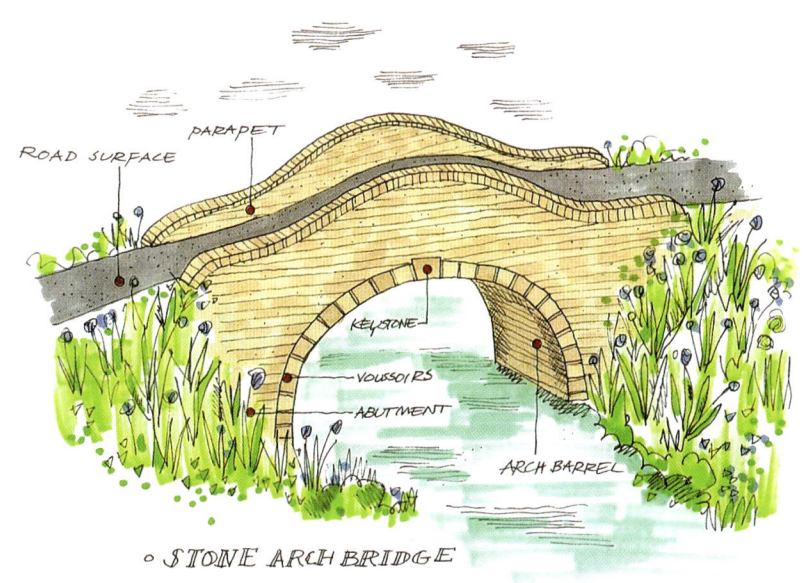

STONE ARCH BRIDGE

the sides, carry the loads across the span. Beam bridges can be single span, or multi-span, or continuous multi-span. The deck spans between the beams; or between transom beams, which themselves span between the beams.

Bearing – The components of a bridge, usually located in the substructure, which bear the weight of the bridge and traffic and transfer the load to the earth. Bearings commonly include abutment walls, piers, arches, and wing walls.

Bedrock – A layer of solid rock that exists at some depth below the soil and gravel above it. Foundations of large structures should rest on bedrock.

Belt Course – A moulding or projecting course of brick or stone running horizontally along the face of a bridge placed to divert rainwater from the face of the bridge. These may sometimes be ornate to provide decoration. Also known as a 'string' course.

Bridge Deck – The roadway or surface that carries traffic.

C

Cable-stayed – A bridge form in which the deck of the roadway is supported by angled cables connected directly to towers without the use of vertical suspenders.

Camber – This is the gradual slope of the roadway, or along the length of the bridge. Camber is needed to ensure that a bridge deck drains to the sides (or to one side) and does not pond water; there is also a matter of appearance: the bridge deck looks better if it has a slight upward curve along the length, even though this does not affect the strength or durability of the bridge.

Cantilever – Any horizontal structural member that is securely fastened only at one end and suspends out over open space at the other.

Capping – Stones or masonry that is placed on the top of a wall or parapet to protect it from damage by the elements. Also called coping stones.

Capstone – The top or finishing stone of a masonry structure such as a pier, wall, or plinth.

Carriageway – That part of the bridge deck that includes all traffic lanes, hard shoulders, and pedestrian walkways. The carriageway width is the distance between raised kerbs or parapets if there are any.

Cast Iron – A metal alloy that contains 2 to 4 percent carbon and whose shape is cast in a mould. It is brittle, non-malleable and cannot be bent.

Catenary – The uniform curve formed by a flexible wire suspended at both ends. In suspension bridges the catenary curve is formed by the primary suspension cables from which the vertical suspenders are hung.

Causeway – A raised path or road across wet ground or water.

Chamfer – A cut that is made in a material, usually as a 45-degree angle, to produce a flat sloping surface.

Channel Arch – A bridge arch that spans the channel of a river and through which water continuously or usually flows.

Cladding – Non-structural surface coverings attached to the outside of a building or bridge to protect it from weather or for aesthetic reasons.

Coade Stone – First marketed at the turn of the 1770s, Coade stone was a remarkable new building material. Using a recipe that was not fully understood until the 1990s, its makers claimed to have produced the first ever 'artificial stone'. Tough and hard-wearing, it offered new opportunities for fine-detailed decoration. Just as extraordinary as the stone was the person who sold it: Eleanor Coade, one of the few women to be acknowledged as a major influence on 18th-century architecture.

Cofferdam – A watertight enclosure built within a body of water (river) from which water is pumped to expose the riverbed and permit the dry construction of supporting piers.

Concrete – A hard composite material first used extensively by the Romans. It is made by mixing fine aggregates such as gravel and sand with cement and water. In addition to its hardness, its great advantage is that it can be poured into a multitude of shapes.

Coping – The sloping top course of a wall or parapet designed to repel rainwater and protect the structure below.

Corbel – An architectural support consisting of a stone or small shelf projecting out from a wall to support a weight superimposed upon it. A masonry bracket.

Cornice – A horizontal moulding that projects out from a wall to protect the wall face or to crown it aesthetically.

Counterweight – Any object or weight that is intended to equal another weight and provide balance.

Course – A continuous horizontal layer of brick or stone bonded with mortar. Many courses together form a wall.

Cross Bracing – Structural support members that intersect to form an 'X' and that transfer weight or pressure.

Cross-section – A virtual view made possible by a cut plane across the axis of the member, structure, or any construction.

Crown – The highest point of the extrados of an arch. The 'keystone' will always sit at the crown of an arch.

Culvert – A closed conduit or tunnel used to convey water from one area to another, normally from one side of a road to the other side. Typically, culverts are box shaped, round, or elliptical in cross-section. They are often prefabricated and can be made from pipes, reinforced concrete, or other materials that are embedded within the surrounding landscape.

Cutwater – The triangular shaped wedge located on the upstream piers of a bridge pier, that resists (cuts) the flow of water and ice and prevents scouring.

D

Deck – The superstructure component of a bridge that carries the traffic. See also 'Carriageway'.

Designation – The recognition of particular heritage value(s) of a significant place by giving it formal status under law or policy intended to sustain those values.

Diaphragm (Transverse) – A member that resists lateral forces and transfers loads to support. The main function of diaphragms is to provide stiffening effect to deck.

E

Eaves – The edge of a roof structure that protrudes out from the structure's walls.

Extrados – The outer curve or surface of an arch or vault. The opposite of 'Intrados'.

F

Fabric – The material substance of which bridges are formed, including construction material, geology, archaeological deposits, structures and buildings, and flora.

Fabrication – To construct or put together something (bridge) made from semi-finished or raw materials.

Falsework – Temporary wooden scaffolding or support structures used in arched bridge construction to hold the components (voussoirs and vaulting) in place until the construction is sufficiently advanced to support itself. See also 'Timber Centring'.

Flood Arch – The arched openings in a bridge or viaduct over a valley or lowland through which floodwater may pass during high water. These are used in distinction from a channel arch, which spans the channel and through which water continuously or usually flows. Also distinct from a culvert, which is a tube and has a closed bottom side.

Footbridge – A structure designed to carry non-vehicular traffic such as pedestrians, over a span.

Footing – The base course of concrete or stone at the bottom of a wall or construction that is designed to spread its weight and support the load of the structure above.

Foundation – A stone or concrete load bearing part of a structure that supports a building from underneath, below ground. The term is often used interchangeably with footing.

G

Girder – An important horizontal structural member or large beam of steel or concrete that supports smaller components of a bridge such as the roadway.

Grade I – For bridges, Grade I (one) indicates that the structure is of 'exceptional interest'.

Grade II – For bridges, Grade II (two) indicates that structures are of 'special interest', warranting every effort to preserve them.

Grade II* – For bridges, Grade II* (two star) indicates that structures are 'particularly important…of more than special interest'.

H

Heritage Bridge – Any bridge which makes an important and significant contribution to the heritage of the country in terms of cultural, historic, design and engineering, or political.

Historic England – Officially 'The Historic Buildings and Monuments Commission for England', more commonly known as Historic England, established in 1984 by the National Heritage Act of 1983. Historic England is the Government's advisory body on the historic environment in England.

Historic Interest – A classification by Historic England. To be of special historic interest a bridge must illustrate important aspects of the nation's social, economic, cultural, or military history, and/or have close historical associations with nationally important people. There should normally be some quality of interest in the physical fabric of the structure itself to justify the statutory protection afforded by this classification. *(Historic England, 2018)*

I

I-Beam – A rolled steel or concrete girder with a cross section resembling an 'I'.

Impost – The 'springing' point of an arch in an abutment or pier. Where the bottom voussoir rests.

Intrados – The interior curve or surface of an arch or vault.

Iron – An abundant ferrous metal that is strongly attracted by a magnet and has the elemental symbol of Fe. It is very strong and used in making steel. Other forms used in bridge construction are cast iron and wrought iron. It also appears naturally in stone and some historical bridges are constructed of ironstone.

Ironwork – Articles or parts of a bridge that are made of iron, especially if made in a decorated way.

K

Keystone – The wedge-shaped voussoir at the crown of an arch that locks the other stones in place. The keystone is often more decorative than the other voussoirs.

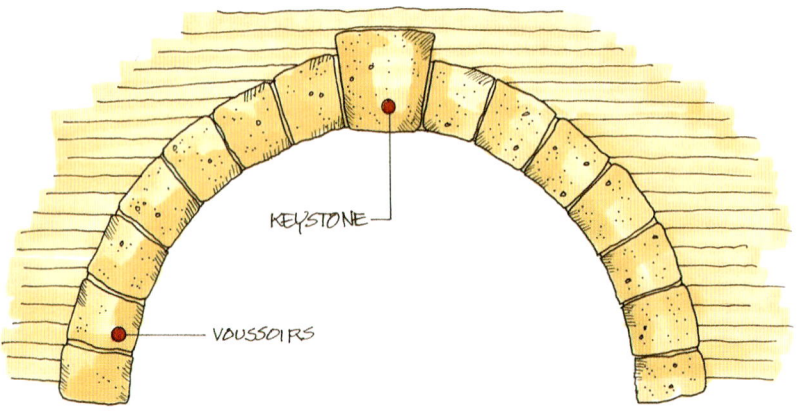

KEYSTONE

VOUSSOIRS

KEYSTONE

L

Lattice Girder – A truss girder where the upper and lower flanges are connected by intersecting diagonal bars forming a lattice of structural support for the load.

Limestone – A white or light grey sedimentary rock consisting mainly of calcium carbonate that is used as an abundant building material and in the making of cement.

Listed Structure – Any structure that is included in a list compiled or approved by the Secretary of State including:

(a) any object or structure fixed to the bridge.

(b) any object or structure within the curtilage of the bridge, which, although not fixed to the building, forms part of the land and has done so since before 1 July 1948.

Load – The weight of bridge components and traffic. Load may be either live (moving) load or dead (stationary) load. The maximum amount of weight any given bridge can carry is called its load limit. The live load limit of a bridge will be greater than its dead load limit.

Lug – Any kind of a projection for carrying or supporting something.

M

Masonry – The craft of building in stone, clay, brick, or concrete. The term may also refer to the finished work of a mason such as a masonry bridge.

Member – Any individual structural component.

Monument – Monument may be defined as:

(a) any building, structure, or work, whether above or below the surface of the land, and any cave or excavation; or

(b) any site comprising the remains of any such building, structure, or work or of any cave or excavation; and

(c) any site comprising, or comprising the remains of, any vehicle, vessel, aircraft or other movable structure or part thereof which neither constitutes nor forms part of any work which is a monument within paragraph (a) above; and any machinery attached to a monument shall be regarded as part of the monument if it could not be detached without being dismantled. *(Historic England, 2018)*

Mortar – A thick paste like mixture of sand, water, and cement or lime which is put between bricks and stone to hold them together. Cement based mortar becomes hard when it dries.

Moulding – A shaped strip of masonry that projects out from the bridge face. These may be either for decorative purposes or for use as a drip course.

N

National Importance – The Secretary of State may on first compiling the Schedule of Monuments include therein any monument which appears to him/her to be of national importance.

P

Palladian – A classical style of architecture based on the works of Andrea Palladio (1508–1580). It is based on the symmetry, perspective, and values of the formal classical temple architecture of the Ancient Greeks and Romans.

Parapet – The upper portions of a wall or bridge face that protect pedestrians from falling. These may be solid wall like structures or balustrades.

Pedestrian Bridge – Any bridge designed to carry pedestrians and non-vehicular traffic across an impediment such as a river, canal, or railway.

Pediment – A triangle shape used to adorn Grecian architecture. A pediment is often used to form small gables and triangular decorations over niches, doors, and windows.

Pier – The supporting structures built into the ground between the abutments, in the gap, which carry intermediate spans of a multi-span bridge.

Pile – Long lengths of timber, steel, or reinforced concrete that are driven lengthwise into the ground to act as foundational support for a structure above.

Plantsman – A horticulturalist who grows, studies, collects, and supplies plants to others. A person skilled with and knowledgeable about plants.

Plinth – A slab-like base or platform that is used to support something above. Bridge piers rest on plinths built into the riverbed using cofferdams.

Prefabricate – To manufacture sections of a bridge or building offsite so that they can be shipped and put together quickly.

Projecting Course – A course of masonry that is set out further from the face of a wall or bridge face. These may be either for decorative purposes or for use as a drip course.

Q

Quatrefoil – A French word meaning four leaves. A conventionalised representation of a flower with four petals or of a leaf with four leaflets. This is a popular motif in bridge architecture.

Quoin – The blocks or brick forming the corner of a structure. Often these are more visually prominent than the wall material to give a sense of substance or permanence. In bridges quoins are most obvious along the edge of the arch. Voussoirs act as quoins in this instance.

R

Reach – A section of river or stream between a defined upstream and downstream location, for which the flow is measured at a point somewhere along that section. Generally, a reach of a river will be representative of conditions in that section of river or stream and will be uniform with respect to discharge, depth, area, and slope.

Reinforced Concrete – Concrete in which steel rods or mesh have been embedded so that the two materials acting together create a building material that is very resistant to forces of load or stress.

QUATREFOIL

Repair – Work beyond the scope of maintenance, to remedy defects caused by decay, damage, or use, including minor adaptation to achieve a sustainable outcome, but not involving restoration or alteration.

Rib – A long narrow, usually curved, construction element used to add support and strength. An example would be the supporting stone structural elements, 'ribs', in the vaults of arches in medieval bridges.

Rising – The point at which a river's water 'rises' to the surface of the land and is often qualified with an adverbial expression of place. For example: the River Thames rises in Gloucestershire.

Rustication – The act of making something look rustic (unfinished). In architecture it is the type of decorative masonry created by chamfering the edges of stone blocks or tooling worm tracks, vermiculation, in the faces of stone blocks.

S

Sandstone – A sedimentary rock composed of fine grains of compacted quartz sand, felspar, and silica. Sandstone may be any colour, but the most common colours are tan, brown, yellow, red, and pink.

Scheduled Monument – Any monument (bridge) that is included in the schedule which is compiled and maintained by the Secretary of State for Culture, Media and Sport.

Segmental Arch – A segmental arch, which is also known as a Syrian arch, forms a partial curve covering less than 180 degrees. It has a small rise in the centre and is semi-elliptical across the top. It is shorter in height than the semi-circular arch.

Shear – The force acting perpendicular (in cross section) to an object.

Sluice – An artificial passage for water (as in a millstream) fitted with a valve or gate for stopping or regulating flow.

Soffit – The horizontal underside of a structural element such as an arch or overhang.

Spalling – Spalling (sometimes incorrectly called 'spaulding' or 'spalding') is the result of water entering brick, concrete, or natural stone. Freezing and thawing forces the surface to peel, pop out, or flake off. It is also known as 'flaking', especially in limestone.

Span – The clear distance from face to face of supports. The length of any bridge segment between its structural supports. There may be one, two, or several spans within a gap.

Spandrel – The roughly triangular area above and on the sides of the arch of a bridge, running below the string course and above the voussoirs.

Steel – An alloy of iron and small amounts of carbon that has a high tensile strength and yet is malleable. Used extensively as a building material.

Stress – In building materials, the action withing a structural member such as a beam or pier that resists change when an outside force such as weight is acting upon it.

String Course – A moulding or projecting course of brick or stone running horizontally along the face of a bridge placed to divert rainwater from the face of the bridge. These may sometimes be ornate to provide decoration. Also known as a 'belt' course.

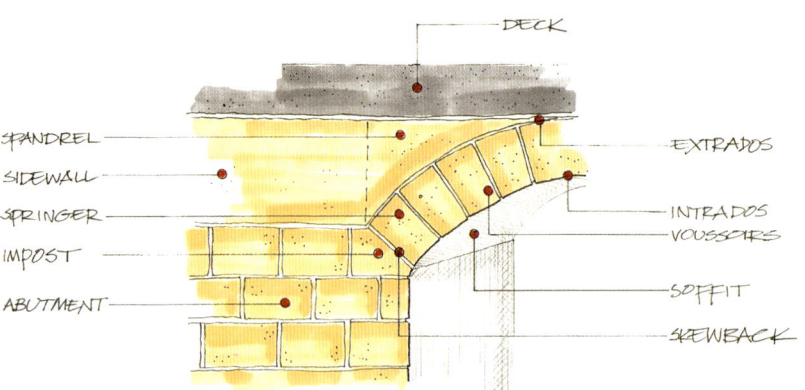

HALF ARCH

Substructure – The components of a bridge hat are below the road deck level and bear the load of the bridge downward into the ground.

Super-incumbent Load – Any load that rests and exerts pressure upon another structural element.

Superstructure – That part of a building or bridge that is carried by the main supporting level such as piers, foundations, and abutments.

Suspenders – In suspension bridges these are the cables or iron rods that hold up the bridge deck.

Suspension Bridge – A suspension bridge has its road deck supported at regular intervals by cables on both sides. These run from anchorages at both ends across two or more towers within the span. The cables are usually made from multi-strand wires bundled together. (A suspension bridge is similar to a steel cable-stayed bridge except that the suspension cables are smooth curves, rather than a number of straight stays.)

Symmetry – Having a balance of shape, proportion, and content. Something is said to have symmetry if it is the same on both halves. The simplest form of symmetry is reflection.

T

Tensile Strength – The strength necessary to resist tensile (pulling) strain. A rope or the suspenders of a bridge have a defined pulling or tensile strength.

Thrust – The outward pressure exerted horizontally or diagonally against an abutment due to the loading being carried by it.

Timber – In building , a term applied to a piece or pieces of sawn lumber. In bridgebuilding it was the first material used because of its availability, strength, and lightweight.

Timber Centring – A wooden framework temporarily built under bridge vaults and arches to sustain them while under construction. It is removed, and possibly re-used, once the arch is complete. See also 'Falsework'.

Transoms – Steel beams that span under the roadway and carry the loads of the roadway to the trusses or beams.

Transverse – Extending at a 90-degree angle from something else. Sideways in relationship.

Truss Bridge – Bridges where trusses carry the loads across the span. The roadway may go through, between the trusses (a through-truss bridge) or over the trusses (an over-truss bridge, sometimes referred to as an under-truss bridge).

V

Vault – The underside of a stone arch. Sometimes also referred to as the 'barrel'.

Vermiculation – Stonework that is decorated in such a way to resemble wormlike tracks. Vermiculation can happen naturally is some stone types or it can be tooled by a mason onto the stone surface.

Viaduct – Any elevated passageway, especially a bridge, that carries traffic over a gorge or ravine. The term is most generally applied to railway bridges across low lying floodplains.

Voussoir – In architecture, any one of the wedge-shaped stones used in forming an arch. The middle one is called the keystone.

W

Web Beam – The central portion of an -beam that connects the flanges and resists shear. In bridges these are often seen in 'Beam' bridges where it is necessary to reduce weight but resist shear. Similar to, but slightly different from, a lattice beam.

Wing Walls – The walls either side of abutments which retain the embankment of the roadway and help prevent erosion.

Wrought Iron – A commercial form of iron almost entirely free of carbon, which is malleable, tough, and yet relatively soft. It can be easily magnetised and bent into shapes.

Bibliography

Amersham Museum, 1908. Amersham Old Town buildings. [Online]
Available at: https://amershammuseum.org/history/old-town/
[Accessed 1 July 2020].

Amersham Museum, 2018. 191 (Town Mill). [Online]
Available at: https://amershammuseum.org/history/old-town/high-street-north/191-town-mill/
[Accessed 5 July 2020].

Baines, A. & Birch, C., 1994. *A Chesham Century: The Story of a Town and it's Council 1894–1994.* London: Baron Birch.

Brimacombe, P., 2001. *Capability Brown: The Master Gardner.* Andover: Pitkin Unichrome.

British History Online – Amersham, 2001. The hundred of Burnham: Amersham. [Online]
Available at: https://www.british-history.ac.uk/vch/bucks/vol3/pp141-155
[Accessed 5 July 2020].

British History Online – Ickford, 2001. Parishes: Ickford. [Online]
Available at: https://www.british-history.ac.uk/vch/bucks/vol4/pp56-61
[Accessed 3 July 2020].

British History Online – Olney, 2001. Parishes: Olney with Warrington. [Online]
Available at: https://www.british-history.ac.uk/vch/bucks/vol4/pp429-439
[Accessed 2 July 2020].

British History Online – Thame, 2001. Thame: Topography, manors and estates. [Online]
Available at: https://www.british-history.ac.uk/vch/oxon/vol7/pp160-178
[Accessed 1 July 2020].

British History Online – Waddesdon, 2001. Parishes: Waddesdon with Westcott and Woodham. [Online]
Available at: https://www.british-history.ac.uk/vch/bucks/vol4/pp107-118
[Accessed 3 July 2020].

British History Online – Buckingham, 2004. The borough of Buckingham. [Online]
Available at: https://www.british-history.ac.uk/vch/bucks/vol3/pp471-489
[Accessed 9 July 2020].

British History Online – Chesham, 2004. Parishes: Chesham. [Online]
Available at: https://www.british-history.ac.uk/vch/bucks/vol3/pp203-218
[Accessed 9 July 2020].

Buckingham Society, 2016. Buckingham's London Road Bridge. [Online]
Available at: http://www.buckinghamsociety.org.uk/newsreports/London%20Road%20Bridge.pdf
[Accessed 17 June 2020].

Buckinghamshire Archives, 2020. Buckinghamshire County Council Photo Archive. [Online]
Available at: https://www.buckscc.gov.uk/services/culture-and-leisure/buckinghamshire-archives/online-resources/historic-photographs/
[Accessed July 2020].

Buckinghamshire County Council, 2017. Sharing Wycombes Old Photographs. [Online]
Available at: http://swop.org.uk/swop/swop.htm
[Accessed 12 June 2020].

Burke, A. E., Dalell, J. R. & Townsend, G., 1950. *Architectural and Building Trades Dictionary.* Chicago (Illinois): American Technical Society.

Concrete Renovations, 2019. What are the differences between Grade I and II listed buildings? [Online]
Available at: https://www.concreterenovations.co.uk/news/what-are-the-differences-between-grade-i-and-ii-listed-buildings/
[Accessed 1 August 2020].

Engineering Timelines, 1999. Marlow Bridge. [Online]
Available at: http://www.engineering-timelines.com/scripts/engineeringItem.asp?id=1493
[Accessed 18 June 2020].

Engineering, Timelines 1999. Tickford Bridge. [Online]
Available at: www.engineering-timelines.com/scripts/engineeringItem.asp?id=1123
[Accessed 10 April 2020].

Engineering Timelines, 2020. Maidenhead Road Bridge. [Online]
Available at: http://www.engineering-timelines.com/scripts/engineeringItem.asp?id=484
[Accessed 1 June 2020].

Farrow, R., 2014. Thame Bridge on the old Thame Road. [Online]
Available at: https://www.geograph.org.uk/photo/3852246
[Accessed 23 June 2020].

Ferguson, S., 2008. Village Road Denham. [Online]
Available at: https://www.geograph.org.uk/photo/849976
[Accessed 28 June 2020].

Geni, 1999. Stowe Landscape Gardens & Monuments, Buckinghamshire, England. [Online]
Available at: https://www.geni.com/projects/Stowe-Landscape-Gardens-Monuments-Buckinghamshire-England/29131
[Accessed 9 July 2020].

Goodland, J., 2019. Turvey Bridge – a brief history. [Online]
Available at: https://www.turveyhistory.org.uk/topics/buildings/turvey-bridge-a-brief-history
[Accessed 19 June 2020].

Harris, M., 1968. Newport Pagnell's Iron Bridge. *Wolverton and District Archeological Journal,* Volume 1, pp. 60–63.

High Wycombe Society, 2019. Pann Mill Watermill. [Online]
Available at: http://www.pannmill.org.uk/
[Accessed 28 June 2020].

Historical Cookham, 2011. First Cookham Toll Bridge. [Online]
Available at: https://widbrook2.blogspot.com/2011/10/first-cookham-tollbridge.html
[Accessed 12 July 2020].

Historical Cookham, 2011. The Second Cookham Toll Bridge. [Online]
Available at: https://widbrook2.blogspot.com/2011/10/second-cookham-tollbridge. html
[Accessed 23 July 2020].

Historic England – Hartwell House, 1997. Hartwell House. [Online]
Available at: https://historicengland.org.uk/listing/the-list/list-entry/1000192
[Accessed 3 August 2020].

Historic England – Chenies Place, 1999. Chenies Place (Woodside). [Online]
Available at: https://historicengland.org.uk/listing/the-list/list-entry/1000594
[Accessed 10 July 2020].

Historic England – Claydon, 1999. Claydon. [Online]
Available at: https://historicengland.org.uk/listing/the-list/list-entry/1000597
[Accessed 3 August 2020].

Historic England – Ickford, 2007. Ickford Bridge together with Whirlpool Arch Bridge. [Online]
Available at: https://historicengland.org.uk/listing/the-list/list-entry/1159729
[Accessed 4 August 2020].

Historic England – Five Arch Bridge, 2016. Five Arch Bridge. [Online]
Available at: https://historicengland.org.uk/listing/the-list/list-entry/1332826
[Accessed 4 August 2020].

Historic England, 2018. Scheduled Monuments. [Online]
Available at: https://historicengland.org.uk/listing/what-is-designation/scheduled-monuments/
[Accessed 4 July 2020].

Historic England – Hughenden, 2019. Hughenden Manor. [Online]
Available at: https://historicengland.org.uk/listing/the-list/list-entry/1000318
[Accessed 18 July 2020].

Historic England, n.d. Tickford Bridge. [Online]
Available at: https://historicengland.org.uk/listing/the-list/list-entry/1125464
[Accessed 17 May 2020].

History of Bridges, 2020. Structure, components and parts of bridge. [Online]
Available at: http://www.historyofbridges.com/facts-about-bridges/bridge-parts/
[Accessed 28 June 2020].

Horn, G., 2013. Misbourne Bridge. [Online]
Available at: https://www.geograph.org.uk/photo/3434810
[Accessed 29 June 2020].

Lutyens Trust, 1999. Miscellaneous works. [Online]
Available at: http://www.lutyenstrustexhibitions.org.uk/miscellaneous-works/4578837599
[Accessed 3 June 2020].

Lyte, C., 1980. *The Thames.* Hove: Wayland Publishers Ltd..

Maidenhead.net, 2000. History. [Online]
Available at: https://www.maidenhead.net/history/
[Accessed 12 March 2020].

Mallett, T., 2012. *Cookham Bridge.* (Oil on canvas) Cookham: Private collection.

Mathews, C., 2013. Bridge over stream, Denham, Buckinghamshire. [Online]
Available at: https://www.geograph.org.uk/photo/3511972
[Accessed 28 June 2020].

Mawrey, G., 2018. An Unsung Arcadia. *Historic Garden Review,* November, Issue 38, pp. 6–10.

MK Heritage, n.d. Old Stratford: the bridge over the River Ouse. [Online]
Available at: http://www.mkheritage.co.uk/os/doc/property/bri.html
[Accessed 14 February 2020].

Myers, D., 2004. *Maidenhead Bridge.* (Oil on canvas) Maidenhead: Private collection.

Myers, D., 2005. *Cookham Bridge.* (Oil on canvas) Maidenhead: Private collection.

Myers, D., 2016. Cookham Bridge. [Online]
Available at: https://thames.me.uk/s00780.htm
[Accessed 18 June 2020].

National Trust, 2016. Capability Brown at Stowe. [Online]
Available at: www.capabilitybrown.org/sites/default/files/
capability_brown_at_stowe.pdf
[Accessed 22 July 2020].

Nether Winchendon House, 2019. The gardens and grounds of Nether
Winchendon House. [Online]
Available at: https://www.nwhouse.co.uk/the-grounds.html
[Accessed 30 July 2020].

Newport Pagnell Historical Society – Heritage Trail, 2020. A brief history of
Newport Pagnell. [Online]
Available at: http://www.mkheritage.org.uk/nphs/a-brief-history-of-newport-
pagnell/
[Accessed 2 July 2020].

Newport Pagnell Historical Society – Heritage Trail, 2020. Newport Pagnell
heritage trail. [Online]
Available at: http://www.mkheritage.org.uk/nphs/heritage-trail-walk-one/
[Accessed 2 July 2020].

North Bucks Times, 1934. Bodies found in Shipton Field. *North Bucks Times,*
17 April.

Northampton Mercury, 1800. Accident on Old Stratford Bridge. [Online]
Northampton Mercury, 25 January, p. 3.
Available at: https://www.britishnewspaperarchive.co.uk/viewer/bl/0000317/
18000125/012/0003

Olney & District Historical Society, 2019. A chronological history of Olney.
[Online]
Available at: https://www.mkheritage.org.uk/odhs/intro-olney-history/
[Accessed 7 July 2020].

Peters, K., 2009. The Battle of Aylesbury was Oliver Cromwell really there?
Aylesbury Town Council Magazine, December, pp. 17–18.

Pevsner, N., 1991. *London 3: North West.* London: Yale University Press.

Phillips, G., 1981. *Thames Crossings: Bridges, Tunnels, and Ferries.*
Newton Abbot: David & Charles.

Piper, J., 1940. *The Bridge, Tyringham.* (Watercolour) London: The Victoria
and Albert Museum.

Ratcliff, O., 1907. *Olney Bucks.* Olney: Cowper Press.

Spencer, S., 1936. *View from Cookham Bridge.* (Oil on canvas)
Cookham: Stanley Spencer Gallery.

Taunt, H., 1875. *Cookham Bridge from North Shore.* (Oil on canvas)
London: Private collection.

The Marlow Society, 2009. Marlow Bridge 1957–1962. [Online]
Available at: http://www.marlowsociety.org.uk/marlow-bridge-1957-
1962/index.php
[Accessed 17 June 2020].

The Turvey Website, 2016. Turvey Bridge. [Online]
Available at: http://www.turveybeds.com/turveybridge.html
[Accessed 19 July 2020].

The Upper & Bedford Ouse Catchment Partnership, 2016. About the Ouzel
& Milton Keynes catchment. [Online]
Available at: https://ubocp.org.uk/catchments/ouzel-and-milton-keynes/
[Accessed 30 June 2020].

Turvey Historical Society, 2016. John Higgins the artist. [Online]
Available at: https://www.turveyhistory.org.uk/catalogue_item/longuet-
higgins-collection/turvey-abbey-commonplace
[Accessed 19 July 2020].

Winslow History, 1999. Shipton. [Online]
Available at: http://www.winslow-history.org.uk/shipton.shtm
[Accessed 4 July 2020].

Wombwell, E. G., 1861. *The North Bridge.* (Oil on canvas) Newport Pagnell:
Newport Pagnell Historical Society.

Acknowledgements

The author would like to gratefully acknowledge the contributions made by the following:

Katherine Gwyn, *Senior Community Outreach and Projects Archivist, Buckinghamshire Archives*

Councillor Jane MacBean, *Buckinghamshire Council – Chiltern Ridges, Chesham Town Council – Lowndes*

Meghan Evans, *Image Researcher, National Trust*

Keith Dolan, *Structures Team Leader, Transport for Buckinghamshire*

Mark Cowie, *Illustrator*

Marion Townsend, *Photographic Coordinator*

Chris Sims, *Graphic Designer*

Peter Hawkes, *Copy Editor & Publisher*

The Staff at:

Amersham Museum, Amersham

Buckingham Old Gaol Museum, Buckingham

Buckinghamshire County Museum, Aylesbury

Chesham Town Library, Chesham

Chesham Historical Society, Chesham

Cowper and Newton Museum, Olney

Maidenhead Heritage Centre, Maidenhead

Marlow Museum, Marlow

Milton Keynes Museum, Wolverton

Newport Pagnell Historical Society and Museum, Newport Pagnell

Old Trails Museum, Winslow

Stanley Spencer Gallery, Cookham

Stowe Gardens, Buckingham

Thame Museum, Thame

West Wycombe Park, West Wycombe

Winslow Historical Society and Museum, Winslow

Wycombe Museum, High Wycombe

Mark Cowie

Mark Cowie is a UK-based illustrator, his work has appeared at the Cartoon Museum in London. He produces hand drawn maps of cities and towns with an ironic twist.

His work often appears under his alias of Tony Hantz.

Email: **cowieandhantz@gmail.com**

Instagram: **@tonyhantz**

Chris Sims

Chris Sims is a graphic designer and fine artist. He has worked for over 30 years in the design industry, producing brochures, advertising material, record sleeves, magazines and books. As an abstract fine artist he is gaining a worldwide reputation with his work being collected across several continents.

Email: **studio384@me.com**

Instagram: **@studio384**

Website: **https://web.marcelforart.com/chris_sims**